Tourism and Travel during the Cold War

The Iron Curtain was not an impenetrable divide, and contacts between East and West took place regularly and on various levels throughout the Cold War. This book explores how the European tourist industry transcended the ideological fault lines and the communist states attracted an ever-increasing number of Western tourists. Based on extensive original research, it examines the ramifications of tourism, from sun-and-sea package tours to human rights travels, in key Eastern European locations including East Berlin, the Soviet Union, Yugoslavia, and Albania. The book's analysis of the politics, culture, and history of tourism to the East offers important new perspectives on European tourism in the twentieth century.

Sune Bechmann Pedersen holds a PhD in History and is Researcher in Media History at Lund University. He is currently working on the project "Holidays behind the Iron Curtain: The Politics of Scandinavian Tourism to Communist Europe, 1945–1989," funded by the Swedish Research Council.

Christian Noack is Associate Professor for East European Studies at the University of Amsterdam. He has published broadly on the history of tourism in the Soviet Union and currently leads a HERA project on the "European Spa as a Public Space and a Social Metaphor."

Routledge Studies in the History of Russia and Eastern Europe

Tourism and Travel during the Cold War

Negotiating Tourist Experiences across the Iron Curtain

Edited by Sune Bechmann Pedersen and Christian Noack

Routledge
Taylor & Francis Group

LONDON AND NEW YORK

First published 2020 by Routledge

2 Park Square, Milton Park, Abingdon, Oxon, OX14 4RN
605 Third Avenue, New York, NY 10017

Routledge is an imprint of the Taylor & Francis Group, an informa business

First issued in paperback 2020

British Library Cataloguing-in-Publication Data
A catalogue record for this book is available from the British Library

Library of Congress Cataloging-in-Publication Data
A catalog record for this book has been requested

ISBN: 978-0-367-19212-9 (hbk)
ISBN: 978-0-367-77727-2 (pbk)

Typeset in Times New Roman
by Apex CoVantage, LLC

Contents

Illustrations

Abbreviations

AIEST	International Association of Scientific Tourism Experts
Comecon	Council for Mutual Economic Assistance
CSCE	Conference on Security and Cooperation in Europe
DNL	Det Norske Luftfartsselskap (The Norwegian Aviation Company)
EEC	European Economic Community
ETC	European Travel Commission
FRG	Federal Republic of Germany
GDR	German Democratic Republic
IATA	International Air Transport Association
IUOTO	International Union of Official Travel Organisations
MBDC	Mixed Bulgarian–Danish Commission for Economic, Industrial, and Technical Cooperation
NATO	North Atlantic Treaty Organisation
NCSJ	National Conference on Soviet Jewry
OECD	Organisation for Economic Cooperation and Development
OEEC	Organisation for European Economic Cooperation
RFE	Radio Free Europe
SAS	Scandinavian Airlines System
TSJ	Turistički Savez Jugoslavije (The Tourism Association of Yugoslavia)
UNECE	United Nations Economic Commission for Europe
UNESCO	United Nations Educational, Scientific and Cultural Organisation

Notes on contributors

Sune Bechmann Pedersen holds a PhD in History and is Researcher in Media History at Lund University. He is currently working on the project "Holidays behind the Iron Curtain: The Politics of Scandinavian Tourism to Communist Europe, 1945–1989," funded by the Swedish Research Council.

Kathleen Beger received her PhD in History from the University of Regensburg in 2018 for the thesis *Erziehung und "Unerziehung" in der Sowjetunion: Das Pionierlager Artek und die Archangelsker Arbeitskolonie im Vergleich* [Education and "Uneducation" in the Soviet Union: The Pioneer Camp Artek and the Arkhangelsk Labour Colony Compared].

Lonneke Geerlings is an interdisciplinary historian, biographer, editor and organiser. Her PhD research at Vrije Universiteit in Amsterdam focuses on the transnational life of Rosey E. Pool (1905–1971), a Dutch–Jewish translator, educator, and anthologist of African American poetry.

Shaul Kelner is Associate Professor of Sociology and Jewish Studies at Vanderbilt University. He is the author of *Tours That Bind: Diaspora, Pilgrimage, and Israeli Birthright Tourism* (NYU Press, 2010).

Karl Lorentz Kleve is Curator and Head of Research at the Norwegian National Aviation Museum in Bodø, where he has worked for 20 years with the history of Norwegian aviation.

Christian Noack is Associate Professor for East European Studies at the University of Amsterdam. He has published broadly on the history of tourism in the Soviet Union and currently leads a HERA project on the "European Spa as a Public Space and a Social Metaphor."

Angela Romano holds a PhD in International History and is an expert in the Cold War, European integration, and cooperation. She is currently a senior research fellow at the Department of History and Civilisation and Manager of the ERC-funded project *PanEur1970s*.

Michelle Standley holds a PhD in History from New York University. Her research and writing on tourism, cities and globalisation, and contemporary art has appeared in numerous publications, including *Journal of Architecture* and in several edited volumes. She is currently completing work on an illustrated book about immigrant-run shops in New York City.

Elitza Stanoeva holds a PhD in History (Technical University of Berlin, 2013) and is currently Research Associate at the Department of History and Civilisation within the ERC project *PanEur1970s* at the EUI as the team expert on Bulgaria.

Adelina Stefan is a postdoctoral fellow in the Department of History and Civilisation at the European University Institute in Florence, Italy. She holds a PhD in History from the University of Pittsburgh (2016) with a thesis entitled "Vacationing in the Cold War: Foreign Tourists to Socialist Romania and Franco's Spain."

Igor Tchoukarine earned his doctorate in history from the Ecole des hautes études en sciences sociales (EHESS) in Paris. He is currently Lecturer in History and Editor of the Centre for Austrian Studies Newsmagazine at the University of Minnesota.

Francesco Zavatti, PhD, is a post-doctoral researcher at the Institute for Contemporary History, Södertörn University, and lecturer at the Department of Political Science, Roma Tre University.

Acknowledgements

This volume is the result of a series of three conferences and workshops held at the University of Amsterdam and the University of Gothenburg between 2015 and 2017. We are very grateful to Dina Fainberg who initiated the first workshop in March 2015 and to the European Travel Cultures research group of the Amsterdam School for Regional, Transnational and European Studies under whose auspices and with whose financial support the first two meetings took place. At the University of Amsterdam we would also like to thank the many colleagues of the Department of History, European Studies and Religious Studies who chaired sessions and commented on papers. We are particularly grateful to Tatjana Das and Paul Koopman for their untiring logistical support.

A generous grant from the Swedish Research Council allowed us to hone the papers at a third workshop, held at the Centre for European Research of the University of Gothenburg in December 2017. We are most grateful to the Centre for European Research for its support and to Linda Berg, Birgit Karlsson, and Angelica Sohlberg for their exceptional organisational skills. Finally, we would like to thank all the participants of the three events for the lively and engaged discussions.

Crossing the Iron Curtain

An introduction

Sune Bechmann Pedersen and Christian Noack

Tourism – travel in pursuit of pleasure – is an essential ingredient of modernity. In the twentieth century, tourism was democratised and transformed into a mass phenomenon. The right to rest, leisure, and annual paid holidays was a feature of both the 1936 Soviet Constitution and the Universal Declaration of Human Rights of 1948. The three competing societal systems of the twentieth century – capitalism, communism, and fascism – sought to engage their citizens in leisure travel.[1] When the post-Stalinist Soviet Union joined the global contest to provide its citizens with "the good life," tourism became a Cold War battleground.[2] On either side of the East–West divide, tourism proved "too important to leave to the private sector alone."[3]

Focusing on Western tourism behind the Iron Curtain, this volume sheds light on how the post-war European tourist industry challenged and overcame the ideological fault lines and enabled ever-increasing mobility across the Iron Curtain. We analyse the politics and economics of Western tourism in Eastern Europe. Tourism mattered to the socialist bloc's balance of payments and substantiated official claims to "peaceful coexistence," while in the 1970s it was a target of high-level diplomacy during the Conference on Security and Cooperation in Europe (CSCE) process, with several provisions of the Helsinki Final Act dedicated to the question of mobility across the Iron Curtain. We also go beyond an analysis of policies and institutions to explore how individual holidaymakers from various Western countries experienced and made sense of their journeys behind the Iron Curtain. Thus we investigate first, how and why Eastern Europe became a tourist destination for citizens of the West; second what impact this had on the development of a tourism industry in the Eastern bloc; and third to what extent the experiences of Western tourists in Eastern Europe influenced mutual perceptions and Cold War stereotypes of "the other." In so doing, we engage with three major trends and debates in recent historiography: the histories of transnational tourism, the cultural Cold War, and mobilities in the supposedly backward and static societies in Eastern Europe.

The history of tourism has travelled far in the past two decades. It was not long ago that scholars working on the history of travel and tourism habitually lamented their topic's exclusion from the "charmed circle of acceptable themes in European history."[4] Tourism was considered a trivial topic and "the enduring stereotypes of tourists as herdlike, superficial gazers doggedly seeking amusement . . . hampered serious scholarly investigation."[5] Today, however, tourism history is flourishing. The questions of tourism, travel, and mobility are firmly established at the heart of contemporary debate about consumption, migration, globalisation, and climate change. "To study tourism is to study the history of the modern world," as Eric Zuelow recently put it, echoing Dean MacCannell's classic study, *The Tourist*.[6]

Tourism, in the broadly shared Western understanding of the term as leisurely travel, is by definition a modern phenomenon, as it presupposes the genuinely modern concept of leisure time. With industrialisation and the rise of the middle classes, tourism spread across social boundaries and geographical borders. Mac-Cannell's take on modern tourism, ironically modelled after the pre-modern practice of religious pilgrimage, emphasised tourism's role in the spatial and cultural authentication of modern nations and states by visiting "shrines" – landscapes and sights framed as being enduringly significant for a national community.[7] In nineteenth-century Central and Eastern Europe ruled by the German, Habsburg, and Russian empires, the nation-building process coincided with the spread of tourism. As routes, landscapes, and sights were marked or relabelled as national, tourism helped carve out spatial identities, and after 1918 was firmly ingrained in the state-building process.[8]

In comparison, the Soviet state was a late bloomer. While denouncing the allegedly idle emptiness of bourgeois tourist practice, the regime strove to juxtapose an alternative model of purposeful tourism within the boundless Soviet territory. With "proletarian tourists" as trailblazers for forging a new Soviet people, the Soviet Union's external borders served at the same time as spatial markers of the new transnational Soviet community.[9] This reinvention of tourism posed a number of dilemmas for the Soviet leadership as it accentuated the conflict between the purposeful and the pleasurable, austerity and consumption, collective goals and individual desires.[10]

The Sovietisation of Central and Eastern Europe and the Balkans between roughly 1948 and 1956 meant that the idiosyncratic Soviet model of tourism spread to those countries, and with it the inherent dilemmas. How these dilemmas were accentuated as the East European countries, each with its own pre-socialist tourism history, organised their own state-sponsored tourism sectors, was thus contingent on the continuities and breaks with national tradition.[11]

Tourism is by default entangled in competing nationalising and internationalising forces. The authorities were frequently tempted to close their borders for fear of dangerous diseases and the ideas that foreigners might bring. With rallying cries such as "See America first!," "Know your country!" (Sweden), and "Discover your homeland!" (Czechoslovakia), travel promoters around the world sought to democratise and promote domestic tourism by kindling a national consciousness.[12] Meanwhile, the drive to experience the exotic abroad has always compelled travellers to cross geographic and territorial borders. The growing interest in tourism history, closely related to the wider turn to transnational and global history, has confirmed the advantages of looking beyond the nation obvious to historians wary of methodological nationalism.[13] In a time infatuated with connectivity, tourism presents an ideal prism through which to view ideas, people, and commodities as they circulate and move across borders.[14]

Cold War studies have witnessed a similar transformation in the wake of the cultural turn. The history of entanglements, transfers, and transnational connections have broadened a field traditionally preoccupied with international relations conducted by a powerful elite. The Cold War was rooted in ideological antagonism, geographical division, and nuclear deterrence. The latter resulted in a stalemate, especially on the European continent, which made the "cultural front" all the more important as a Cold War battleground. The response, naturally enough, was to study the Cold War competition for hearts and minds around the world.[15] Historians now embrace a plurality of themes and methods in their work, and operate with a pragmatically broad understanding of the field, investigating everything "from alliance diplomacy and political manipulation to development projects, from cultural and intellectual confrontations all the way to bloody 'proxy wars' in allegedly 'peripheral' areas."[16]

More than anything else, the study of cultural competition from 1945 to 1989 has taught us that the Cold War was awash with connections and exchanges across the East–West divide.[17] Rather than an impenetrable Iron Curtain, the Eastern bloc erected a "semipermeable membrane," which permitted select goods, people, and information to pass through.[18] In fact, the barriers that each East European state erected were permeable in ways that often changed several times in the period. Innocuous goods could usually pass in both directions, and the ease with which Westerners could cross the Iron Curtain generally increased over the years. East European citizens were allowed to travel to the West as tourists, too, although in much smaller numbers. Early in the Cold War this was a luxury granted to only the most trusted cadres, and while the Soviet Union upheld this principle until the end, travel policy in other countries followed a circuitous pattern of liberalisation and restriction.[19] In the second half of the 1960s, for instance, a growing number of ordinary Czechoslovaks went on holiday in the West, but this ended when the Normalisation regime tightened the screw again in 1969.[20] The 1970s and 1980s also saw East Europeans emigrate in greater numbers to the West, which added another dimension to the external barriers or membrane.[21] The Iron Curtain was thus perforated with loopholes. It was as much a mental as a physical barrier, and, being constituted by a multitude of actors and actions, was emphatically dynamic.[22]

The democratisation of international tourism in the West during the post-war boom coincided with the post-Stalinist opening of Eastern Europe to the outside world. After a hiatus of about a decade following Sovietisation in 1948, a steadily increasing number of ordinary tourists from North America and Western Europe started to visit the Soviet Union and the other countries behind the Iron Curtain.[23] Western interest in visiting Eastern Europe gave the socialist regimes an opportunity to showcase their societies, and to earn hard currency while doing so. However, Western tourists also presented the risk of espionage, and could be a negative influence on local populations by exposing them to foreign fashions and ideas. In other words, tourism in the Cold War was a field of competing cultural, economic, ideological, and security concerns.

The efforts to attract foreign tourists raised a number of fundamental questions that had significant ideological ramifications. Which attractions should be promoted to Western visitors? Bourgeois pre-socialist destinations such as Bohemia's spas, or the accomplishments of state socialism? How were visitors to reach those destinations and be guided, fed, and accommodated? The various answers to these questions point to the fluctuations in the regimes' self-confidence and the unstable balance between economic and security interests, liberalisation, and periods of intensified social control. Controlling Western tourists, however, became increasingly difficult with the admission of individual travellers with private means of transport.

International visitors competed with domestic travellers for the same, often scarce tourism resources. As recent research on East European mobility have shown, mobility, and private motor tourism in particular, was far more common in the Eastern bloc than conventional wisdom has it.[24] Studies have also focused on voluntary and involuntary migration, finding that both social and spatial mobility was greater than previously assumed.[25] To examine tourism behind the Iron Curtain thus means engaging with important domestic developments in infrastructure, as the socialist states, too, "established frameworks and incentives that people on the move elaborated and constituted in their own ways, often leading to unintended consequences."[26] And all the while the socialist tourism administrations struggled to keep the two groups apart.[27]

Cold War tourism or tourism in the Cold War?

Cold War Europe offered a plurality of real and imagined borders for the tourist to cross. Immediately after the war even the real territorial borders were often imaginary, and long stretches of the border between East and West were poorly demarcated. Tourists and soldiers on leave occasionally strayed into enemy territory by mistake, leading to diplomatic debacles and accusations of espionage.[28] The Iron Curtain that Churchill spoke of in 1946 was not a physical border, but an abstract, ideological divider. As the Cold War progressed, however, the physical East–West border was increasingly fortified with barbed wire, roadblocks, and watchtowers.

The ideological conflict gave Westerners' travel in Eastern Europe an exotic flavour, akin to the kind of holidays in sites of suffering, disaster, and death usually described as dark tourism.[29] Some Western companies made a point of offering tours with a Cold War flavour, for instance to Yugoslavia, "the country next to the Iron Curtain."[30] Western guidebooks to Eastern Europe, and especially to Berlin, the frontline of Cold War Europe, reflected the changing intensity of the conflict and its potential for dark tourism. Fielding's 1951 edition of *Travel Guide to Europe* called Berlin "a keg of dynamite and definitely *not* recommended under any circumstances."[31] A Swedish guidebook from 1973 anticipated its readers' Cold War–induced fears by asking "Is it dangerous to travel to Berlin?" although it answered calmly in the negative.[32] In 1988, Fodor's guide to Eastern Europe comforted the reader that "travel to Eastern Europe is by no means the uncertain, complex affair it once was, and all the Eastern bloc countries are extremely eager to attract Western visitors."[33]

The Cold War and communism could indeed serve as attractions in their own right, as the contributions to this volume by Michelle Standley and Shaul Kelner illustrate. Standley's study of bus tours to East Berlin in the 1970s points to the desire of West German tourists to have their impression of a backward East Germany confirmed. Kelner's analysis shows that many American tourists pictured themselves as Cold War undercover agents when they visited Soviet Jews in the 1970s and 1980s. In both cases, Cold War mentalities clearly coloured the tourist experience of East Berlin and Moscow. The mingling of leisure culture and international politics is reminiscent of Americans' "Cold War holidays" in France studied by Christopher Endy.[34]

It is important to stress, though, that the Cold War was by no means the only lens through which tourists viewed this part of Europe under communist rule. There is an important distinction to be made between Cold War tourism and tourism in the Cold War. While the Cold War was at the centre of the first form of tourism, it was a contextual chronological marker in the second.[35] In some parts of Eastern Europe, the pre-socialist tourist legacy survived the advent of a socialist regime. For example, the spas of Bohemia and Budapest had long been established destinations for health tourists from all over Europe, and they continued to attract some Western guests after the Second World War.[36] Having survived the war comparatively unscathed, socialist Czechoslovakia and Hungary also appealed to a cultured middle class who wanted to see the spectacular architecture of Central Europe.[37] To consider Western tourism in Eastern Europe solely in terms of Cold War conflict and capitalist–socialist competition is to risk overlooking the tourist traditions that the regimes were sometimes happy to perpetuate. Moreover, as the chapters by Elitza Stanoeva, Adelina Stefan, and Igor Tchoukarine show, the South East European tourist authorities worked hard to overcome the barriers to Western tourism raised by the Cold War.

The distinction between Cold War tourism and tourism in the Cold War raises the question of how the key processes and caesuras of Cold War history and tourism history mapped onto one another. The contributions to this volume point to clear links between the periodisation of Cold War historiography and tourism history. The post-Stalinist Soviet Union's attempt to shed its international isolation is the backdrop to Karl Kleve's chapter on the rapidly expanding post-war network of bilateral civil aviation agreements, including between the Scandinavian countries and the USSR. The initiative in 1957 to invite children and youth groups from capitalist countries to spend the summer at the Soviet Artek camp, studied by Kathleen Beger, was part of Nikita Khrushchev's ideas for peaceful cooperation and mutual understanding. Meanwhile, Yugoslavia after the Stalin–Tito split pursued its own plans for opening up to the West, as Igor Tchoukarine shows in his chapter on the country's engagement with the global travel industry in the early Cold War.

Yugoslavia's success with international tourism inspired the tourism sectors in Bulgaria and Romania, studied here by Elitza Stanoeva and Adelina Stefan. The Council for Mutual Economic Assistance (Comecon) first discussed tourism at a summit in 1955, and in 1957 held a conference devoted to tourism. Initially, the focus was international tourism within the socialist bloc, but as the chapters by Stanoeva, Stefan, and Standley show, from the 1960s onwards, the socialist countries increasingly sought to develop international tourism across the East–West divide. The objective of the nascent South East European tourist industry was to extract hard currency from sun-and-sea holidaymakers, even though the more relaxed international relations at the time also meant that adventurous individuals could explore Eastern Europe. However, as Lonneke Geerlings shows in her chapter on Rosey E. Pool's 1965 Trans-Siberian railway journey, going off the beaten track did not necessarily result in a more profound exchange of ideas or experiences than on a Black Sea beach. The linguistic gulf between the Slavic and the Germanic world proved too great even for a polyglot like Pool. For the most committed fellow travellers, such as the Swedish tourists in Albania studied by Francesco Zavatti, an encounter with the realities of their Stalinist utopia often resulted in mutual misunderstandings and a cognitive dissonance that proved difficult to gloss over.

The contributions to this volume thus point to a clear relation between the history of tourism behind the Iron Curtain and the political history of the Cold War in Europe. Key political events such as the Soviet-led invasions of Hungary in 1956, Czechoslovakia in 1968, and Afghanistan in 1979 had an immediate impact on Western tourism in Eastern Europe. In each case, the response to the military campaigns included political pressure to cancel travel, which temporarily reduced the tourist flow from the West. Within a year or two, though, things were usually back to normal.

Tourism behind the Iron Curtain also followed – and became mixed up with – the positive course of European detente and cooperation from the mid-1960s to the end of the Cold War. As a symbol of improved international relations, official tourist organisations around the world declared 1967 the International Tourist Year. In Eastern Europe the occasion was celebrated with a considerable easing of visa regulations. In subsequent years, as Angela Romano shows, tourism was a feature of the CSCE process and the Helsinki Final Act, while human rights activists, encouraged by the pledges made by the signatories to the accords, joined the growing traffic across the Iron Curtain, adding a new dimension to East–West tourism studied here by Shaul Kelner.

The relation between tourism and human rights enshrined in the Universal Declaration and stressed in the Helsinki Final Act raises the question of the role of tourism in the 1989–1991 caesura. The socialist regimes' promise to deliver a better life than in the West generated a rising demand for consumption that the regimes were never able to satisfy. Unflattering comparisons between East and West prompted by tourist encounters increased the strain on the socialist regimes, but domestic and intra-bloc tourism was in fact one of the appreciated pleasures in socialism.[38] To be sure, the complexity of tourism as a "product" laid bare the contradictions and flaws of the socialist system and its economy.[39] But the official sponsorship of tourism and mobility as integral to the socialist project facilitated individual and even subversive appropriations by the population, and thus allowed for "socialist escapes," which may in the long run have had a stabilising effect on socialist societies.[40]

As solidarity between the East European regimes began to falter in 1989, however, tourism did play a part in the ultimate fall of communism. When Hungary relaxed its border controls with Austria in the summer of 1989, East German holidaymakers poured in and stayed on in the hope of escaping to the West. Faced with a humanitarian crisis, Hungary eventually allowed the East Germans free passage to Austria, which forced Czechoslovakia to close its border with Hungary. Thousands of East German "tourists" then besieged the West German embassy in Prague, and they too eventually won passage to the West. The socialist attempt to satisfy citizens' wanderlust by having relatively open borders inside the Eastern bloc thus ultimately accelerated the bloc's disintegration.[41]

The fall of communism resulted in the wholesale re-evaluation of education and work, and former tourism workers were often at an advantage in the new economy.[42] Two chapters in this volume address this Cold War legacy: Kathleen Beger's and Adelina Stefan's oral history interviews with former summer camp pioneers and tourism workers, which show how experience of the tourism sector proved valuable in the new market economy. Interaction with Westerners had given them better language skills, connections in the West, and perhaps even a more entrepreneurial mindset that helped them launch their own businesses and ultimately to navigate post-communism.

The chronologies of the Cold War and of international tourism cannot be understood without also taking the long-term socioeconomic developments into account. The introduction of statutory paid leave in the interwar period and the switch to a five-day working week in the 1950s and 1960s, for example, changed the patterns and concepts of consumption and leisure. In terms of technological development, the spread of private car ownership in the same period and the first wide-body planes were important milestones. All of this happened independently of the Cold War, yet posed particular challenges to the socialist regimes and planned economies of Eastern Europe as they tried to satisfy their citizens' growing demands for consumption and recreation.

In order to question the relationship between Cold War tourism and tourism in the Cold War, we must also reconsider the realities of state security and central planning. The security services on either side of the Iron Curtain harboured serious concerns about the liberalisation of mobility between the blocs. As Karl Kleve shows here, commercial air routes between East and West were met with scepticism because they made spying easier. The socialist security apparatuses also feared that tourism would make it harder to catch Western infiltrators and easier for citizens to escape. If nothing else, greater contact with capitalist lifestyles would give the lie to socialist promises of conscious consumption and purposeful leisure habits.[43] In an attempt to curb the negative effects, vigilance campaigns regularly cautioned the public against interaction with Western tourists, as they might be spies in disguise – in fact, a justified fear as spies from both sides did of course pose as tourists.[44] The important point, though, is that despite the power and influence wielded by the security apparatuses in Eastern Europe, and despite the fears they harboured about Western tourism, the political and economic gains from Western tourism largely overrode their security concerns. The history of tourism in the Cold War was thus one of enduring tension between competing institutional interests inside the socialist leadership and bureaucracy.

The opening up for Western tourists must be understood against this backdrop. As Alex Hazanov reminds us, the opening up of the Soviet Union in the mid-1950s was by no means "a careful experimental process," but rather "a series of ad hoc decisions that together amounted to a profound historical shift."[45] Although more research is still needed on this process in other East European countries, there is little evidence to suggest that the process was any different there. Western tourism required the involvement of a multitude of different organisations, with conflicting interests resulting in improvised and often-revised policies.[46]

Historiography has made tremendous progress since the publication in 1991 of Derek Hall's seminal volume *Tourism and Economic Development in Eastern Europe and the Soviet Union*.[47] At that point, the authors were forced to draw on personal observation to supplement problematic official statistics and tendentious publications.[48] Since then, the opening of the government archives – East and West – has provided historians with a far more nuanced understanding of the regimes. The contributors to this volume can thus rely on solid archival evidence when studying the evolution of the tourism sector and its economic and political significance at home and abroad.

Tourism historians generally find that institutional developments and the textual and visual representations of destinations are better documented in the archives than the experiences of individual tourists. Here, however, historians of the socialist regimes of Eastern Europe are at an advantage, because of the excruciating detail in which the regimes monitored and policed the sentiments of their citizens and visiting foreigners. True, secret police files are often unreliable and must be used with great care. Yet when triangulated with oral history interviews and other sources, they provide unique insights into foreign visitors' experiences and their encounters with locals. In addition to the official sources, the contributors to this volume draw on diaries, letters, memoirs, travel accounts, and contemporary news reports held in private and institutional archives. The various combinations of sources and methods employed by the contributors together provide a complex and nuanced picture of tourism and travel behind the Iron Curtain.

Outline of the book

Recent studies have pointed to the abundance of capitalist thinking to be found in the socialist economic systems.[49] The first section of the book focuses on the development of a socialist tourism industry, with contributions by Stanoeva and Stefan that show the commercial concerns that drove the South East European tourism industry. Influenced by neighbouring Yugoslavia and Greece, Bulgaria and Romania took foreign tourism seriously as a motor of economic development. Earlier than the other countries in the socialist bloc, for which they would be models, Bulgaria and Romania specifically addressed a Western European clientele in the new market for package holidays.

Every socialist country had at least one agency that dealt specifically with foreign tourists. In most cases it was the responsibility of the state tourist agencies such as the USSR's Intourist, Bulgaria's Balkantourist, and Romania's ONT Carpaţi. The primary objective of these institutions was to earn hard currency, and they all opted for collaboration with commercial tour operators in the West in their bid for a share of the leisure market.[50] Their ability to offer package tours to foreigners generally depended on a plethora of other organisations, from the trade unions' domestic tourism infrastructure to transportation networks to the catering industry. In most cases, this prevented the emergence of full-blown tourist industries with ministerial backing, capable of channelling the necessary resources into the development of domestic and international tourism. One exception to this rule, as Elitza Stanoeva shows, was Bulgaria, which had comparatively little heavy industry, so the tourist sector had less competition in the institutional hierarchy.

The presence of Western tourists was fraught with unintended consequences – as Stefan shows, it enabled smuggling networks and black markets – yet "the lure of capitalism" still prevailed over the negative effects. The single-party states found it difficult to suppress encounters and exchanges between Western tourists and citizens involved in the service infrastructure. Personal encounters between travellers and hosts were often accompanied by the exchange of personal services for cash or commodities, which was officially forbidden but almost impossible to prevent. This led to tacit acquiescence and attempts by governments to cream off the profits in the shape of foreign currency shops and the like.

Michelle Standley's chapter, by contrast, deals with East German efforts to use mass tourism to "market" socialism with the help of guided bus tours to East Berlin. Not that this was without its attempts to earn hard currency with the help of the *Zwangsumtausch*, the compulsory exchange of Western D-Marks for much less valuable Ostmarks, but the main aim was to impress the day trippers with the achievements of state socialism in the divided city. However, Standley analyses how the official East German tourist agency, Reisebüro-DDR, largely failed to make a dent in the convictions of its customers. Day trippers either completely ignored all attempts to mediate the "socialist experience," preferring to aggressively reiterate Western stereotypes and narratives, or engaged very selectively with the tour guides' narratives. The East German attempts to anticipate and tweak the tourists' perceptions failed miserably.

Perception management posed less of a problem in the case of the package tours to Bulgaria and Romania, where Western commercial companies partnered with Eastern agencies to promote what were essentially cheap sun-and-sand holidays on the other side of the Iron Curtain. Package tours were ostensibly non-political, and the idea of the "Gold Coast" as a beach holiday paradise did not necessarily corroborate the Cold War narrative of the drab life under communism.

The tensions between possible exposure to propaganda and the individual pursuit of tourist experiences are central to the second section of the book. Kathleen Beger's chapter on the famous soviet holiday camp for Young Pioneers, Artek, illustrates the negotiations between visitors and hosts as to the meaning of Soviet internationalism. Artek celebrated a very specific amalgam of national cultures in a Soviet-style social straitjacket that nevertheless left room for individual agency. The need for mediation and translation between East and West produced a new type of international (youth) tourism specialist, who was sometimes capable of building a career on the experiences gathered in Artek, as Beger's interviews show.

Drawing on ego-documents, Lonneke Geerling's analysis of a visit by two experienced women activists and travellers to the Soviet Union in the 1960s exposes both the possible gains and limits of such sources for tourism history. As left-leaning feminists, Rosey E. Pool and Ursel Isenburg at first glance fit the category of fellow travellers, and they were indeed partisan, their account surprisingly uncritical of their private and public encounters on the trip. At the same time, Pool's contemporary notes are as sketchy as they are startling, providing us with an image of Khrushchev's USSR that owed little to other travelogues of the time. It requires familiarity with the protagonists' lives and personal circumstances to interpret them, as they are a very specific, subjective take and commentary on Western political issues, such as women's liberation or racism in the American South.

Zavatti's chapter likewise problematises fellow travellers. Drawing on press material, friendship society archives, and a curiously multivocal travel account, he shows that Swedish friends of Albania framed their holidays there not exclusively in political terms, but were as much enchanted by the sun, sea, and sand – and possibly sex. As Zavatti shows, travelling through Hoxa's Albania, Swedish tourists did not completely forfeit their agency, although it would seem that their room for manoeuvre was narrower than in the contemporary Soviet Union.

The tourist experience is central also to Shaul Kelner's chapter. Exploring a unique body of written reports produced by American tourists in Brezhnev's Soviet Union in the 1970s, Kelner introduces us to the opposite of a fellow traveller. His human rights travellers were on self-appointed subversive missions behind the Iron Curtain. In this case, the Cold War was not just a temporal framework, but their very *raison d'être*. Cold War clichés about spying and communist repression determined what they assumed they would find and what they hoped to achieve. In the Soviet Union, their expectations were confirmed by the warm welcome that awaited them in the vulnerable privacy of the dissidents' flats, as opposed to the cold official hospitality of Intourist's making.

The chapters in the final section of the book return to the larger question of the economic and political roles which tourism was expected to play in the second half of the twentieth century. Set in the context of the Marshall Plan's support for tourism, Igor Tchoukarine's chapter documents how widespread was the hope that international tourism would trigger political change and economic recovery after the Second World War. In the course of de-Stalinisation, the optimistic view of tourism harboured by Yugoslavia and the new tourism-promoting organisations such as the International Union of Official Tourist Organisations and the European Travel Commission spilled over to the rest of socialist Eastern Europe.[51] Khrushchev was one of the most optimistic if erratic proponents of a "tourist turn" in Soviet policies.[52] Karl Kleve, though, in his account of the framework agreements for international airlines to use Soviet airspace, confirms that the Soviets could muster considerable resistance to opening up "hard" Cold War borders. Air travel in northern Europe, Kleve asserts, was directly affected by Khrushchev's Thaw, but concessions were always a quid pro quo for putting the Soviet carrier Aeroflot on a par with the flag carrier Scandinavian Airlines System (SAS). Where the Soviet Union was unlikely to profit from opening its airspace, as in the question of overflight rights over Siberia for routes to the Far East, such concessions were very slow to materialise.

Did Cold War travel behind the Iron Curtain, then, confirm or frustrate the planners and politicians' hopes that were paradigmatic of the post-war period? Any answer will have to be as inconclusive as the debate over whether East–West encounters facilitated or slowed the collapse of communism. In economic terms, smaller and less developed socialist bloc economies such as Bulgaria profited most from investment in the tourism sector. In most socialist countries, though, tourism remained marginal, and subordinate to other, often ideologically motivated preferences. At the same time, tourism, at least the domestic kind, rose to prominence as part of the consumer turn in state socialism following Stalin's death. With all the necessary differentiation between the Eastern bloc countries, tourism as a very specific economic product never sat easily with the hierarchical and bureaucratic procedures of central planning. On the one hand, this prevented a rise of service industries to prominence in the socialist countries. On the other hand, it required ideological compromises to let it work in practice.[53]

In political terms, tourism behind the Iron Curtain sparked skirmishes in the cultural Cold War. It carried the weight of expectation of both sides, as was evident in the CSCE negotiations of the tourism, visa, and currency regulations in the run-up to the Helsinki Final Act of 1975. The closing chapter by Angela Romano illustrates the optimism of Western politicians about the potential for change that such contacts might mean, epitomised in Egon Bahr's famous rationale for German Ostpolitik, "Wandel durch Annäherung." Eastern bloc leaders happily paid lip service to mutual understanding through increased mobility, but their insistence on control and securitisation repeatedly betrayed their insecurities about tourism's functioning behind the Iron Curtain.

Then again, tourism was too important to simply curtail, in particular at times of detente. The obvious contradictions that international tourists encountered just added to the long list of real-socialist oddities that people in the Eastern bloc had learned to live with, and created openings for the subversive appropriation of such encounters by the tourists themselves.

Notes

1　Shelley Baranowski, *Strength through Joy: Consumerism and Mass Tourism in the Third Reich* (Cambridge: Cambridge University Press, 2007); Diane P. Koenker, *Club Red: Vacation Travel and the Soviet Dream* (Ithaca: Cornell University Press, 2013); Victoria De Grazia, *The Culture of Consent: Mass Organisation of Leisure in Fascist Italy* (Cambridge: Cambridge University Press, 2002 [1981]); Ellen Furlough,

"Making Mass Vacations: Tourism and Consumer Culture in France, 1930s to 1970s," *Comparative Studies in Society and History* 40, no. 2 (1998); Greg Richards, "Politics of National Tourism Policy in Britain," *Leisure Studies* 14, no. 3 (1995).

2 Shelley Baranowski and Ellen Furlough, "Introduction," in *Being Elsewhere: Tourism, Consumer Culture, and Identity in Modern Europe and North America*, ed. Shelley Baranowski and Ellen Furlough (Ann Arbor: University of Michigan Press, 2001), 20.

3 Anne E. Gorsuch and Diane P. Koenker, "Introduction," in *Turizm: The Russian and East European Tourist under Capitalism and Socialism*, ed. Anne E. Gorsuch and Diane P. Koenker (Ithaca: Cornell University Press, 2006), 3.

4 John K. Walton, "Taking the History of Tourism Seriously," *European History Quarterly* 27, no. 4 (1997): 563.

5 Baranowski and Furlough, "Introduction," 1–2.

6 Eric G. E. Zuelow, "Editor's Introduction," *Journal of Tourism History* 10, no. 2 (2018): 103; Dean MacCannell, *The Tourist: A New Theory of the Leisure Class* (Berkeley & Los Angeles: University of California Press, 2013 [1976]).

7 Ibid.

8 Alexander Vari, "From Friends of Nature to Tourist-Soldiers: Nation Building and Tourism in Hungary, 1873–1914," in *Turizm: The Russian and East European Tourist under Capitalism and Socialism*, ed. Anne E. Gorsuch and Diane P. Koenker (Ithaca: Cornell University Press, 2006); Aldis Purs, "'One Breath for Every Two Strides': The State's Attempt to Construct Tourism and Identity in Interwar Latvia," ibid.; Peter Stachel and Martina Thomsen, eds., *Zwischen Exotik und Vertrautem: Zum Tourismus in der Habsburgermonarchie und ihren Nachfolgestaaten* (Bielefeld: Transcript, 2014); Felix Jeschke, "Iron Landscapes: Nation-Building and the Railways in Czechoslovakia, 1918–1938" (PhD diss., University College London, 2016).

9 Evgeny Dobrenko, "The Art of Social Navigation: The Cultural Topography of the Stalin Era," in *The Landscape of Stalinism: The Art and Ideology of Soviet Space*, ed. Evgeny Dobrenko and Eric Naiman (Seattle: University of Washington Press, 2003); Diane P. Koenker, "The Proletarian Tourist in the 1930s: Between Mass Excursion and Mass Escape," in *Turizm: The Russian and East European Tourist under Capitalism and Socialism*, ed. Anne E. Gorsuch and Diane P. Koenker (Ithaca: Cornell University Press, 2006); Christian Noack, "'A Mighty Weapon in the Class War': Proletarian Values, Tourism and Mass Mobilisation in Stalin's Time," *Journal of Modern European History* 10, no. 2 (2012).

10 Koenker, *Club Red*.

11 Adam T. Rosenbaum, "Leisure Travel and Real Existing Socialism: New Research on Tourism in the Soviet Union and Communist Eastern Europe," *Journal of Tourism History* 7, no. 1–2 (2015).

12 Orvar Löfgren, "Know Your Country: A Comparative Perspective on Tourism and Nation Building in Sweden," in *Being Elsewhere: Tourism, Consumer Culture, and Identity in Modern Europe and North America*, ed. Shelley Baranowski and Ellen Furlough (Ann Arbor: University of Michigan Press, 2001); Marguerite S. Shaffer, "'See America First': Re-Envisioning Nation and Region through Western Tourism," *Pacific Historical Review* 65, no. 4 (1996); Felix Jeschke, "'Mountain Men' on 'Iron Horses': National Space in the Representations of New Railway Lines in Interwar Czechoslovakia," *Bohemia* 56, no. 2 (2016): 440.

13 Eric G. E. Zuelow, ed., *Touring Beyond the Nation: A Transnational Approach to European Tourism History* (Farnham: Ashgate, 2011).

14 Sebastian Conrad, *What Is Global History?* (Princeton: Princeton University Press, 2016), 9; Heinz-Gerhard Haupt and Jürgen Kocka, eds., *Comparative and Transnational History: Central European Approaches and New Perspectives* (New York: Berghahn, 2009).

15 Walter L. Hixson, *Parting the Curtain: Propaganda, Culture and the Cold War, 1945–1961* (New York: St Martin's Griffin, 1997); Kiril Tomoff, *Virtuosi Abroad: Soviet*

Music and Imperial Competition During the Early Cold War, 1945–1958 (Ithaca: Cornell University Press, 2015); Sarah Davies, "The Soft Power of Anglia: British Cold War Cultural Diplomacy in the USSR," *Contemporary British History* 27, no. 3 (2013); Peter Romijn, Giles Scott-Smith, and Joes Segal, eds., *Divided Dreamworlds? The Cultural Cold War in East and West* (Amsterdam: Amsterdam University Press, 2012); Patrick Major and Rana Mitter, eds., *Across the Blocs: Cold War Cultural and Social History* (London: Frank Cass, 2004); David Caute, *The Dancer Defects: The Struggle for Cultural Supremacy During the Cold War* (Oxford: Oxford University Press, 2003); Frances Stonor Saunders, *Who Paid the Piper? The CIA and the Cultural Cold War* (London: Granta, 1999).

16 Federico Romero, "Cold War Historiography at the Crossroads," *Cold War History* 14, no. 4 (2014): 689.

17 György Péteri, "Nylon Curtain – Transnational and Transsystemic Tendencies in the Cultural Life of State-Socialist Russia and East-Central Europe," *Slavonica* 10, no. 2 (2004); Simo Mikkonen and Pia Koivunen, eds., *Beyond the Divide: Entangled Histories of Cold War Europe* (New York: Berghahn, 2015).

18 Michael David-Fox, "The Iron Curtain as Semipermeable Membrane: Origins and Demise of the Stalinist Superiority Complex," in *Cold War Crossings: International Travel and Exchange across the Soviet Bloc, 1940s–1960s*, ed. Patryk Babiracki and Kenyon Zimmer (College Station: Texas A&M University Press, 2014), 18.

19 Anne E. Gorsuch, *All This Is Your World: Soviet Tourism at Home and Abroad after Stalin* (Oxford: Oxford University Press, 2011), 17–19.

20 Paulina Bren, "Mirror, Mirror, on the Wall . . . Is the West the Fairest of Them All? Czechoslovak Normalization and Its (Dis) Contents," *Kritika: Explorations in Russian and Eurasian History* 9, no. 4 (2008).

21 Tara Zahra, *The Great Departure: Mass Migration from Eastern Europe and the Making of the Free World* (New York: W. W. Norton, 2016), 250–70.

22 Edith Sheffer, *Burned Bridge: How East and West Germans Made the Iron Curtain* (New York: Oxford University Press, 2011), 4–6.

23 As the methodologies used varied by country and period, historical tourists figures are notoriously difficult to compare. Under communism, tourist figures often drove institutional competition for resources, which further reduce their reliability. With these caveats in mind, though, rough figures for the rapid increase in tourists in the early Cold War are, for the USSR: in 1956 some 500,000 foreign visitors, rising to 1.3 million in 1965, 3.7 million in 1975, and 5 million in 1980. A. D. Popov, " 'Uvidet. Poniat. Poliubit': Sovetskii inostrannyi turizm v kontekste publichnoi diplomatii perioda kholodnoi voiny," *Modern History of Russia* 7, no. 4 (2017): 150. See also Denise Cambau, "Travel by Westerners to Eastern Europe," *ITA Bulletin* 40 (1976).

24 They have been partly inspired what came to be called the "new mobilities paradigm," which suggest a cross-fertilization between the historical sub-disciplines of transport history, migration history, and mobilities studies in general. Colin G. Pooley, *Mobility, Migration and Transport: Historical Perspectives* (Cham: Palgrave Macmillan, 2017), 1–20. See also "Connecting Historical Studies of Transport, Mobility and Migration," *Journal of Transport History* 38, no. 2 (2017). Luminita Gatejel, "Driving Behind the Iron Curtain: Automobility in the Eastern Bloc," *Mobility in History* 7, no. 1 (2016); *Warten, hoffen und endlich fahren: Auto und Sozialismus in der Sowjetunion, in Rumänien und der DDR (1956–1989/91)* (Frankfurt am Main: Campus Verlag, 2014); Lewis H. Siegelbaum, ed., *The Socialist Car: Automobility in the Eastern Bloc* (Ithaca: Cornell University Press, 2013); Christian Noack, "Brezhnev's 'Little Freedoms': Tourism, Individuality, and Mobility in the Late Soviet Period," in *Reconsidering Stagnation in the Brezhnev Era: Ideology and Exchange*, ed. Dina Fainberg and Artemy M. Kalinovsky (Lanham: Lexington Books, 2016).

25 John Randolph and Eugene M. Avrutin, eds., *Russia in Motion: Cultures of Human Mobility since 1850* (Urbana: University of Illinois Press, 2012); Kathy Burrell and

Kathrin Hörschelmann, eds., *Mobilities in Socialist and Post-Socialist States: Societies on the Move* (Basingstoke: Palgrave Macmillan, 2014); Ralf Roth and Henry Jacolin, eds., *Eastern European Railways in Transition: Nineteenth to Twenty-First Centuries* (Farnham: Ashgate, 2013).

26 Charles Steinwedel, "Governing Mobility: Preface," in *Russia in Motion: Cultures of Human Mobility since 1850*, ed. John Randolph and Eugene M. Avrutin (Urbana: University of Illinois Press, 2012).

27 Kristen Ghodsee, *The Red Riviera: Gender, Tourism, and Postsocialism on the Black Sea* (Durham: Duke University Press, 2005), 92–98.

28 See, for example, the week 35 report from the US Embassy, Prague, to the Department of State, Washington, 30 October 1959, 749.00(w)/10–3059. Records of the Department of State relating to Internal Affairs: Czechoslovakia 1955–59. Archives Unbound, Gale CENGAGE Learning.

29 Note though that the concept of dark tourism remains a problematic term for several reasons as pointed out by Tony Seaton, "Encountering Engineered and Orchestrated Remembrance: A Situational Model of Dark Tourism and Its History," in *The Palgrave Handbook of Dark Tourism Studies*, ed. Philip R. Stone et al. (London: Palgrave Macmillan, 2018).

30 Sune Bechmann Pedersen, "A Paradise Behind the Curtain: Selling Eastern Escapes to Scandinavians," in *Eden für Jeden? Touristische Sehnsuchtsorte in Mittel- und Osteuropa von 1945 bis zur Gegenwart*, ed. Bianca Hoenig and Hannah Wadle (Göttingen: V&R unipress, 2019), 245.

31 Temple Fielding, *Fielding's Travel Guide to Europe* (New York: William Sloane Associates, 1951), 359.

32 Gert Hammerby, *Turen går till Berlin* (Stockholm: Geber, 1973), 3.

33 Richard Moore, ed., *Fodor's 89 Eastern Europe* (New York: Fodor's Travel Publications, 1988), 11.

34 Christopher Endy, *Cold War Holidays: American Tourism in France* (Chapel Hill: University of North Carolina Press, 2004).

35 For a discussion of the Cold War as an analytical concept and a chronological marker, see Joel Isaac and Duncan Bell, "Introduction," in *Uncertain Empire: American History and the Idea of the Cold War*, ed. Joel Isaac and Duncan Bell (Oxford: Oxford University Press, 2012).

36 David Clay Large, *The Grand Spas of Central Europe: A History of Intrigue, Politics, Art, and Healing* (Lanham: Rowman & Littlefield, 2015).

37 Bechmann Pedersen, "Paradise Behind the Curtain."

38 David Crowley and Susan E. Reid, eds., *Pleasures in Socialism: Leisure and Luxury in the Eastern Bloc* (Evanston: Northwestern University Press, 2010); Anne E. Gorsuch and Diane P. Koenker, "Introduction: The Socialist 1960s in Global Perspective," in *The Socialist Sixties: Crossing Borders in the Second World*, ed. Anne E. Gorsuch and Diane P. Koenker (Bloomington: Indiana University Press, 2013); Mark Keck-Szajbel, "The Borders of Friendship: Transnational Travel and Tourism in the East Bloc, 1972–1989" (PhD diss., University of California, Berkeley, 2013).

39 On the complexity of tourism as a product, see Christoph Hennig, *Reiselust: Touristen, Tourismus und Urlaubskultur* (Frankfurt am Main: Insel, 1997), 159–64.

40 Cathleen M. Giustino, Catherine J. Plum, and Alexander Vari, eds., *Socialist Escapes: Breaking Away from Ideology and Everyday Routine in Eastern Europe, 1945–1989* (New York: Berghahn, 2013).

41 Baranowski and Furlough, "Introduction," 20; Victor Sebestyen, *Revolution 1989: The Fall of the Soviet Empire* (London: Weidenfeld & Nicolson, 2009), 311–24.

42 Ghodsee, *Red Riviera*, 101–3.

43 Michelle Standley, "The Cold War, Mass Tourism and the Drive to Meet World Standards at East Berlin's T.V. Tower Information Center," in *Touring Beyond the Nation: A Transnational Approach to European Tourism History*, ed. Eric G.E. Zuelow

(Farnham: Ashgate, 2011), 221–22; Dragoş Petrescu, "Closely Watched Tourism: The Securitate as Warden of Transnational Encounters, 1967–9," *Journal of Contemporary History* 50, no. 2 (2015).

44 Sune Bechmann Pedersen, "Eastbound Tourism in the Cold War: The History of the Swedish Communist Travel Agency Folkturist," *Journal of Tourism History* 10, no. 2 (2018).

45 Alex Hazanov, "Porous Empire: Foreign Visitors and the Post-Stalin Soviet State" (PhD diss., University of Pennsylvania, 2016), 4.

46 Examples of Central and South East European case studies include Pavel Mücke, *Šťastnou cestu . . .?! Proměny politik cestování a cestovního ruchu v Československu za časů studené války (1945–1989)* (Pelhřimov: Nová tiskárna Pelhřimov, 2017), 158–213; Igor Tchoukarine, "Yugoslavia's Open-Door Policy and Global Tourism in the 1950s and 1960s," *East European Politics & Societies* 29, no. 1 (2015).

47 Derek R. Hall, ed., *Tourism and Economic Development in Eastern Europe and the Soviet Union* (London: Belhaven Press, 1991).

48 For earlier examples of studies seeking to account for the development of tourism in Cold War Eastern Europe, see *Osttourismus – Chance der Begegnung: Urlaubsreisen nach Osteuropa*. Conference, 23–25 October 1978, Loccumer Protokolle 18/1978; Martin A. Garay, *Le Tourisme dans les démocraties populaires européennes*, vols. 3609–3610, Notes et etudes documentaires (Paris: La documentation française, 1969).

49 Katalin Miklóssy and Melanie Ilič, eds., *Competition in Socialist Society* (New York: Routledge, 2014); Suvi Kansikas, *Socialist Countries Face the European Community: Soviet-Bloc Controversies over East – West Trade* (Frankfurt am Main: Peter Lang, 2014).

50 Bechmann Pedersen, "Eastbound Tourism"; Shawn Salmon, "Marketing Socialism: Inturist in the Late 1950s and Early 1960s," in *Turizm: The Russian and East European Tourist under Capitalism and Socialism*, ed. Anne E. Gorsuch and Diane P. Koenker (Ithaca: Cornell University Press, 2006); Andrei Kozovoi, "The Way to a Man's Heart: How the Soviet Travel Agency 'Sputnik' Struggled to Feed Western Tourists," *Journal of Tourism History* 6, no. 1 (2014); Duncan Light, "'A Medium of Revolutionary Propaganda': The State and Tourism Policy in the Romanian People's Republic, 1947–1965," *Journal of Tourism History* 5, no. 2 (2013); Shawn Conelly Salmon, "To the Land of the Future: A History of Intourist and Travel to the Soviet Union, 1929–1991" (PhD diss., University of California, Berkeley, 2008).

51 Frank Schipper, Igor Tchoukarine, and Sune Bechmann Pedersen, *The History of the European Travel Commission, 1948–2018* (Brussels: The European Travel Commission, 2018).

52 On Khrushchev's excitement about the potential of tourism to promote peace and friendship, see Johanna Conterio, "'Our Black Sea Coast': The Sovietization of the Black Sea Littoral under Khrushchev and the Problem of Overdevelopment," *Kritika: Explorations in Russian and Eurasian History* 19, no. 2 (2018): 327–29.

53 Christian Noack, "Coping with the Tourist: Planned and 'Wild' Mass Tourism on the Soviet Black Sea Coast," in *Turizm: The Russian and East European Tourist under Capitalism and Socialism*, ed. Anne E. Gorsuch and Diane P. Koenker (Ithaca: Cornell University Press, 2006).

References

Argenbright, Robert. "Lethal Mobilities: Bodies and Lice on Soviet Railroads, 1918–1922." *Journal of Transport History* 29, no. 2 (2008): 259–76.

Baranowski, Shelley. *Strength through Joy: Consumerism and Mass Tourism in the Third Reich*. Cambridge: Cambridge University Press, 2007.

Baranowski, Shelley, and Ellen Furlough. "Introduction." In *Being Elsewhere: Tourism, Consumer Culture, and Identity in Modern Europe and North America*, edited by Shelley Baranowski and Ellen Furlough, 1–31. Ann Arbor: University of Michigan Press, 2001.

Bechmann Pedersen, Sune. "Eastbound Tourism in the Cold War: The History of the Swedish Communist Travel Agency Folkturist." *Journal of Tourism History* 10, no. 2 (2018): 130–45.

———. "A Paradise Behind the Curtain: Selling Eastern Escapes to Scandinavians." In *Eden für jeden? Touristische Sehnsuchtsorte in Mittel- und Osteuropa von 1945 bis zur Gegenwart,* edited by Bianca Hoenig and Hannah Wadle, 227–49. Göttingen: V&R unipress.

Bren, Paulina. "Mirror, Mirror, on the Wall . . . Is the West the Fairest of Them All? Czechoslovak Normalization and Its (Dis) Contents." *Kritika: Explorations in Russian & Eurasian History* 9, no. 4 (2008): 831–54.

Burrell, Kathy, and Kathrin Hörschelmann, eds. *Mobilities in Socialist and Post-Socialist States: Societies on the Move.* Basingstoke: Palgrave Macmillan, 2014.

Busch, Tracy Nichols. " 'Comrades, Start Your Engines!' Mobility, Legitimacy, and Roads to Socialism in the Soviet Interwar Period." *Canadian – American Slavic Studies* 47, no. 2 (2013): 221–46.

Cambau, Denise. "Travel by Westerners to Eastern Europe." *ITA Bulletin* 40 (1976).

Caute, David. *The Dancer Defects: The Struggle for Cultural Supremacy During the Cold War.* Oxford: Oxford University Press, 2003.

Conrad, Sebastian. *What Is Global History?* Princeton: Princeton University Press, 2016.

Conterio, Johanna. " 'Our Black Sea Coast': The Sovietization of the Black Sea Littoral under Khrushchev and the Problem of Overdevelopment." *Kritika: Explorations in Russian & Eurasian History* 19, no. 2 (2018): 327–61.

Crowley, David, and Susan E. Reid, eds. *Pleasures in Socialism: Leisure and Luxury in the Eastern Bloc.* Evanston: Northwestern University Press, 2010.

David-Fox, Michael. "The Iron Curtain as Semipermeable Membrane: Origins and Demise of the Stalinist Superiority Complex." In *Cold War Crossings: International Travel and Exchange across the Soviet Bloc, 1940s – 1960s,* edited by Patryk Babiracki and Kenyon Zimmer, 14–39. College Station: Texas A&M University Press, 2014.

Davies, Sarah. "The Soft Power of Anglia: British Cold War Cultural Diplomacy in the USSR." *Contemporary British History* 27, no. 3 (2013): 297–323.

De Grazia, Victoria. *The Culture of Consent: Mass Organisation of Leisure in Fascist Italy.* Cambridge: Cambridge University Press, 2002 [1981].

Dobrenko, Evgeny. "The Art of Social Navigation: The Cultural Topography of the Stalin Era." In *The Landscape of Stalinism: The Art and Ideology of Soviet Space,* edited by Evgeny Dobrenko and Eric Naiman, 163–200. Seattle: University of Washington Press, 2003.

Endy, Christopher. *Cold War Holidays: American Tourism in France.* Chapel Hill: University of North Carolina Press, 2004.

Fielding, Temple. *Fielding's Travel Guide to Europe.* New York: William Sloane Associates, 1951 [1948].

Furlough, Ellen. "Making Mass Vacations: Tourism and Consumer Culture in France, 1930s to 1970s." *Comparative Studies in Society & History* 40, no. 2 (1998): 247–86.

Garay, Martin A. *Le Tourisme dans les démocraties populaires européennes.* Notes et études documentaires, vols. 3609–3610. Paris: La documentation française, 1969.

Gatejel, Luminita. *Warten, hoffen und endlich fahren: Auto und Sozialismus in der Sowjetunion, in Rumänien und der DDR (1956–1989/91).* Frankfurt am Main: Campus Verlag, 2014.

———. "Driving Behind the Iron Curtain: Automobility in the Eastern Bloc." *Mobility in History* 7, no. 1 (2016): 117–22.

Ghodsee, Kristen. *The Red Riviera: Gender, Tourism, and Postsocialism on the Black Sea.* Durham: Duke University Press, 2005.

Giustino, Cathleen M., Catherine J. Plum, and Alexander Vari, eds. *Socialist Escapes: Breaking Away from Ideology and Everyday Routine in Eastern Europe, 1945–1989*. New York: Berghahn, 2013.

Gorsuch, Anne E. *All This Is Your World: Soviet Tourism at Home and Abroad after Stalin*. Oxford: Oxford University Press, 2011.

Gorsuch, Anne E., and Diane P. Koenker. "Introduction." In *Turizm: The Russian and East European Tourist under Capitalism and Socialism*, edited by Anne E. Gorsuch and Diane P. Koenker, 1–14. Ithaca: Cornell University Press, 2006.

———. "Introduction: The Socialist 1960s in Global Perspective." In *The Socialist Sixties: Crossing Borders in the Second World*, edited by Anne E. Gorsuch and Diane P. Koenker, 1–21. Bloomington: Indiana University Press, 2013.

Hall, Derek R., ed. *Tourism and Economic Development in Eastern Europe and the Soviet Union*. London: Belhaven Press, 1991.

Hammerby, Gert. *Turen går till Berlin*. Stockholm: Geber, 1973.

Haupt, Heinz-Gerhard, and Jürgen Kocka, eds. *Comparative and Transnational History: Central European Approaches and New Perspectives*. New York: Berghahn, 2009.

Hazanov, Alex. "Porous Empire: Foreign Visitors and the Post-Stalin Soviet State." PhD diss., University of Pennsylvania, 2016.

Hennig, Christoph. *Reiselust: Touristen, Tourismus und Urlaubskultur*. Frankfurt am Main: Insel, 1997.

Hixson, Walter L. *Parting the Curtain: Propaganda, Culture and the Cold War, 1945–1961*. New York: St Martin's Griffin, 1997.

Isaac, Joel, and Duncan Bell. "Introduction." In *Uncertain Empire: American History and the Idea of the Cold War*, edited by Joel Isaac and Duncan Bell, 3–16. Oxford: Oxford University Press, 2012.

Jeschke, Felix. "Iron Landscapes: Nation-Building and the Railways in Czechoslovakia, 1918–1938." PhD diss., University College London, 2016.

———. "'Mountain Men' on 'Iron Horses': National Space in the Representations of New Railway Lines in Interwar Czechoslovakia." *Bohemia* 56, no. 2 (2016): 437–55.

Kansikas, Suvi. *Socialist Countries Face the European Community: Soviet Bloc Controversies over East – West Trade*. Frankfurt am Main: Peter Lang, 2014.

Keck-Szajbel, Mark. "The Borders of Friendship: Transnational Travel and Tourism in the East Bloc, 1972–1989." PhD diss., University of California, Berkeley, 2013.

Koenker, Diane P. "The Proletarian Tourist in the 1930s: Between Mass Excursion and Mass Escape." In *Turizm: The Russian and East European Tourist under Capitalism and Socialism*, edited by Anne E. Gorsuch and Diane P. Koenker, 119–40. Ithaca: Cornell University Press, 2006.

———. *Club Red: Vacation Travel and the Soviet Dream*. Ithaca: Cornell University Press, 2013.

Kozovoi, Andrei. "The Way to a Man's Heart: How the Soviet Travel Agency 'Sputnik' Struggled to Feed Western Tourists." *Journal of Tourism History* 6, no. 1 (2014): 57–73.

Large, David Clay. *The Grand Spas of Central Europe: A History of Intrigue, Politics, Art, and Healing*. Lanham: Rowman & Littlefield, 2015.

Light, Duncan. "'A Medium of Revolutionary Propaganda': The State and Tourism Policy in the Romanian People's Republic, 1947–1965." *Journal of Tourism History* 5, no. 2 (2013): 185–200.

Löfgren, Orvar. "Know Your Country: A Comparative Perspective on Tourism and Nation Building in Sweden." In *Being Elsewhere: Tourism, Consumer Culture, and Identity in Modern Europe and North America*, edited by Shelley Baranowski and Ellen Furlough, 137–54. Ann Arbor: University of Michigan Press, 2001.

MacCannell, Dean. *The Tourist: A New Theory of the Leisure Class*. Berkeley & Los Angeles: University of California Press, 2013 [1976].

Major, Patrick, and Rana Mitter, eds. *Across the Blocs: Cold War Cultural and Social History*. London: Frank Cass, 2004.

Mikkonen, Simo, and Pia Koivunen, eds. *Beyond the Divide: Entangled Histories of Cold War Europe*. New York: Berghahn, 2015.

Miklóssy, Katalin, and Melanie Ilič, eds. *Competition in Socialist Society*. New York: Routledge, 2014.

Moore, Richard, ed. *Fodor's 89 Eastern Europe*. New York: Fodor's Travel Publications, 1988.

Mücke, Pavel. *Šťastnou cestu…?! Proměny politik cestování a cestovního ruchu v Československu za časů studené války (1945–1989)*. Pelhřimov: Nová tiskárna Pelhřimov, 2017.

Noack, Christian. "Coping with the Tourist: Planned and 'Wild' Mass Tourism on the Soviet Black Sea Coast." In *Turizm: The Russian and East European Tourist under Capitalism and Socialism*, edited by Anne E. Gorsuch and Diane P. Koenker, 281–304. Ithaca: Cornell University Press, 2006.

———. "'A Mighty Weapon in the Class War': Proletarian Values, Tourism and Mass Mobilisation in Stalin's Time." *Journal of Modern European History* 10, no. 2 (2012): 231–54.

———. "Brezhnev's 'Little Freedoms': Tourism, Individuality, and Mobility in the Late Soviet Period." In *Reconsidering Stagnation in the Brezhnev Era: Ideology and Exchange*, edited by Dina Fainberg and Artemy M. Kalinovsky, 59–76. Lanham: Lexington Books, 2016.

Osttourismus – Chance der Begegnung: Urlaubsreisen nach Osteuropa. Conference, 23–25 October 1978, Loccumer Protokolle 18/1978.

Péteri, György. "Nylon Curtain – Transnational and Transsystemic Tendencies in the Cultural Life of State-Socialist Russia and East-Central Europe." *Slavonica* 10, no. 2 (2004): 113–23.

Petrescu, Dragoş. "Closely Watched Tourism: The Securitate as Warden of Transnational Encounters, 1967–9." *Journal of Contemporary History* 50, no. 2 (2015): 337–53.

Pooley, Colin G. "Connecting Historical Studies of Transport, Mobility and Migration." *Journal of Transport History* 38, no. 2 (2017): 251–59.

———. *Mobility, Migration and Transport: Historical Perspectives*. Cham: Palgrave Macmillan, 2017.

Popov, A. D. "'Uvidet. Poniat. Poliubit': Sovetskii inostrannyi turizm v kontekste publichnoi diplomatii perioda kholodnoi voiny." *Modern History of Russia* 7, no. 4 (2017): 148–60.

Purs, Aldis. "'One Breath for Every Two Strides': The State's Attempt to Construct Tourism and Identity in Interwar Latvia." In *Turizm: The Russian and East European Tourist under Capitalism and Socialism*, edited by Anne E. Gorsuch and Diane P. Koenker, 97–115. Ithaca: Cornell University Press, 2006.

Randolph, John, and Eugene M. Avrutin, eds. *Russia in Motion: Cultures of Human Mobility since 1850*. Urbana: University of Illinois Press, 2012.

Richards, Greg. "Politics of National Tourism Policy in Britain." *Leisure Studies* 14, no. 3 (1995): 153–73.

Romero, Federico. "Cold War Historiography at the Crossroads." *Cold War History* 14, no. 4 (2014): 685–703.

Romijn, Peter, Giles Scott-Smith, and Joes Segal, eds. *Divided Dreamworlds? The Cultural Cold War in East and West*. Amsterdam: Amsterdam University Press, 2012.

Rosenbaum, Adam T. "Leisure Travel and Real Existing Socialism: New Research on Tourism in the Soviet Union and Communist Eastern Europe." *Journal of Tourism History* 7, nos. 1–2 (2015): 157–76.

Roth, Ralf, and Henry Jacolin, eds. *Eastern European Railways in Transition: Nineteenth to Twenty-First Centuries*. Farnham: Ashgate, 2013.

Salmon, Shawn. "Marketing Socialism: Intourist in the Late 1950s and Early 1960s." In *Turizm: The Russian and East European Tourist under Capitalism and Socialism*, edited by Anne E. Gorsuch and Diane P. Koenker, 186–204. Ithaca: Cornell University Press, 2006.

Salmon, Shawn Conelly. "To the Land of the Future: A History of Intourist and Travel to the Soviet Union, 1929–1991." PhD diss., University of California, Berkeley, 2008.

Saunders, Frances Stonor. *Who Paid the Piper? The CIA and the Cultural Cold War*. London: Granta, 1999.

Schipper, Frank, Igor Tchoukarine, and Sune Bechmann Pedersen. *The History of the European Travel Commission, 1948–2018*. Brussels: European Travel Commission, 2018.

Seaton, Tony. "Encountering Engineered and Orchestrated Remembrance: A Situational Model of Dark Tourism and Its History." In *The Palgrave Handbook of Dark Tourism Studies*, edited by Philip R. Stone, Rudi Hartmann, A. V. Seaton, Richard Sharpley and Leanne White, 9–31. London: Palgrave Macmillan, 2018.

Sebestyen, Victor. *Revolution 1989: The Fall of the Soviet Empire*. London: Weidenfeld & Nicolson, 2009.

Shaffer, Marguerite S. "'See America First': Re-Envisioning Nation and Region through Western Tourism." *Pacific Historical Review* 65, no. 4 (1996): 559–81.

Sheffer, Edith. *Burned Bridge: How East and West Germans Made the Iron Curtain*. New York: Oxford University Press, 2011.

Siegelbaum, Lewis H., ed. *The Socialist Car: Automobility in the Eastern Bloc*. Ithaca: Cornell University Press, 2013.

Stachel, Peter, and Martina Thomsen, eds. *Zwischen Exotik und Vertrautem: Zum Tourismus in der Habsburgermonarchie und ihren Nachfolgestaaten*. Bielefeld: Transcript, 2014.

Standley, Michelle. "The Cold War, Mass Tourism and the Drive to Meet World Standards at East Berlin's T.V. Tower Information Center." In *Touring Beyond the Nation: A Transnational Approach to European Tourism History*, edited by Eric G. E. Zuelow, 215–39. Farnham: Ashgate, 2011.

Steinwedel, Charles. "Governing Mobility: Preface." In *Russia in Motion: Cultures of Human Mobility since 1850*, edited by John Randolph and Eugene M. Avrutin, 19–24. Urbana: University of Illinois Press, 2012.

Tchoukarine, Igor. "Yugoslavia's Open-Door Policy and Global Tourism in the 1950s and 1960s." *East European Politics & Societies* 29, no. 1 (2015): 168–88.

Tomoff, Kiril. *Virtuosi Abroad: Soviet Music and Imperial Competition During the Early Cold War, 1945–1958*. Ithaca: Cornell University Press, 2015.

Vari, Alexander. "From Friends of Nature to Tourist-Soldiers: Nation Building and Tourism in Hungary, 1873–1914." In *Turizm: The Russian and East European Tourist under Capitalism and Socialism*, edited by Anne E. Gorsuch and Diane P. Koenker, 64–81. Ithaca: Cornell University Press, 2006.

Walton, John K. "Taking the History of Tourism Seriously." *European History Quarterly* 27, no. 4 (1997): 563–71.

Zahra, Tara. *The Great Departure: Mass Migration from Eastern Europe and the Making of the Free World*. New York: W. W. Norton, 2016.

Zuelow, Eric G. E. "Editor's Introduction." *Journal of Tourism History* 10, no. 2 (2018): 103–4.

———, ed. *Touring Beyond the Nation: A Transnational Approach to European Tourism History*. Farnham: Ashgate, 2011.

Part I

Organising Western tourism in the East

1 Exporting holidays

Bulgarian tourism in the Scandinavian market in the 1960s and 1970s

Elitza Stanoeva

In the second half of the 1950s, socialist Bulgaria sought to join the new global business of mass tourism.[1] Although lacking a comprehensive strategy, a long-term plan, or even a coherent concept of tourism, the ambition to develop a tourist hub with international outreach and large hard-currency profits was there from the start. In pursuit of this ambition, both tourism's infrastructure and institutional network mushroomed, and by the mid-1960s the contribution of international tourism to the Bulgarian economy was acknowledged to be indispensable. Hosting capitalist tourists who paid with hard cash was the only way to compensate, at least in part, for the trade deficits Bulgaria incurred vis-à-vis their home countries.

Investment in recreational facilities first appeared on the agenda of the Bulgarian Communist Party (BCP) in the mid-1950s, under two different motivations. Providing holiday leisure for the masses was integral to the social reforms in the aftermath of de-Stalinisation, under the new ideological slogan of "satisfying the growing material needs" of Bulgarian workers.[2] Yet, the subsequent boom in resort construction was also prompted by economic pragmatism, and was geared towards meeting the demands of a foreign clientele. Reflecting these heterogeneous incentives, government strategy from the onset differentiated "economic tourism" from "social tourism."[3] While the latter was a socialist welfare service for the domestic public, economic tourism predominantly targeted foreign visitors from the socialist bloc and beyond.[4]

Shaped by divergent operational logics and administered by separate agencies, facilities for the two categories of holidaymakers – high-paying foreigners and state-subsidised locals – developed side by side, yet apart from one another. The rift, both social and spatial, became unbridgeable as domestic and international holiday travel grew, tourism evolved into a distinct sector of the Bulgarian economy, and the administration of its assets was compartmentalised between multiplying organisations. The agencies in charge of modern holiday facilities gradually withdrew from service provision for locals to focus on what their management successfully promoted as "exports in situ," a form of foreign trade that not only complemented the traditional sort, but yielded higher profits and substantially improved the country's balance of payments.[5] Emulating the operation of foreign-trade organisations, Bulgarian tourist enterprises diversified their activities beyond resort development and hotel management into all kinds of business partnerships abroad.

In foreign trade, Bulgaria lagged behind the other socialist regimes in Central and Eastern Europe in gaining access to the Western markets, but its international tourism fared much better in the new global mass tourism industry.[6] Keeping pace with worldwide developments, Bulgarian tourism succeeded not only in placing the country on the European tourist map, but also in inventing itself from scratch as an economic sector catering to the new phenomenon of mass tourism.

This chapter discusses the institutional development of Bulgaria's international tourism as a peculiar variant of a foreign-trade operation, and examines its intrinsic advantages and limitations compared to commodity exchange.[7] I analyse the development of international tourism in Bulgaria from the mid-1950s onwards, and particularly its positioning in the institutional landscape of the socialist economy, to show how commercial tourism advanced from the margins of the national economy (where it suffered from a precarious dependence on the output of more privileged sectors) to its centre, as an economic sector in its own right, entitled to higher standards and priority supply from the country's heavy and light industries. To illustrate this development, I focus on Bulgaria's economic cooperation with Scandinavia. International tourism being one pillar in Bulgaria's economic cooperation with the West in general, it pays to examine tourist deals with the Scandinavian countries in order to highlight the different ways in which the entanglement of tourism, trade, and diplomacy played out in the changing Cold War context.

The birth of Bulgaria's international tourism: Balkantourist in the late 1950s

In 1966, shortly before the start of the summer season, the Research Institute of Radio Free Europe (RFE) began a lengthy report on "Vacationland Bulgaria" with an unexpected geographical leap:

> One would think there was very little in common in the way of statistics between the United States and Bulgaria. Quite the contrary, for, in the field of tourism, both countries entertained over one million foreign tourists during 1965. The comparison ends here, however, for Bulgaria has, for many years, been a country isolated and virtually inaccessible to the rest of the world. . . . But things are very different now. Bulgaria, during the past two years, has made a volte-face and has energetically entered the competition for Europe's tourists, especially for those from the West carrying wallets-full of foreign currency.[8]

Bulgaria had hit the 1 million visitors mark, with the number of Westerners for the first time exceeding that of guests from other Soviet-bloc countries. As surprising as this might have been for Western commentators, for the Bulgarian government it signified a decade of painstaking work to develop international tourism finally bearing fruit. In 1966 – the year when RFE threw the spotlight on Bulgaria's thriving holiday business – the national tourism administration, hitherto a low-ranking directorate under the Council of Ministers, was upgraded as the

Committee on Tourism (CT), an autonomous branch of government. With this institutional reform, tourism gained the legal status of a separate sector of the state economy; its management had aspired to such recognition ever since it was permitted to operate internationally.[9] Although the CT was also put in charge of domestic "social tourism," it was its international activities that raised its legislative standing and shaped its subsequent policies. International tourism and its increasing differentiation from domestic "social tourism," moreover, was what had spurred the sector's growth in the preceding decade.

In the early post-Stalinist years, the business of international tourism was largely delegated to Balkantourist, a small enterprise set up in January 1948 as the successor to a travel bureau that had specialised in the sale of international train tickets before the Second World War. Given the negligible number of tourists to Bulgaria in the immediate post-war period, the enterprise barely stayed afloat, and its operations dwindled because of obsolescence and its troubles maintaining its properties.[10] In 1954, a state decree authorised Balkantourist to sign contracts with foreign travel agencies, including companies in capitalist countries.[11] Since its reestablishment back in 1948, Balkantourist had been affiliated to the Ministry of Foreign Trade (MFT), and once it resumed business it was only natural to profile itself as a peculiar type of a foreign-trade organisation.

To market Bulgaria abroad as a holiday destination, the government focused on the Black Sea coast, taking its cue from Mediterranean countries such as Yugoslavia and Italy, while simultaneously aspiring to compete with what they offered.[12] Seaside resort development in Bulgaria started modestly with the modernisation of an old resort near the port city of Varna. Previously known as "St Constantine and St Elena," it was now renamed Varna Resort – and later again Druzhba (friendship). At the end of 1956, the holiday complex had seven hotels with just over 1,000 beds; a year later, when investment had increased threefold, ten new hotels and 500 accommodation units described as "little cottages (tents)" were being built to welcome another 2,268 guests.[13]

To gauge whether the investment had paid off and the resort warranted further development, the government commissioned the Central Statistical Directorate to evaluate Varna Resort's revenue from foreign holidaymakers, which it did by meticulously calculating the hard-currency income relative to domestic-currency costs. It used indicators modelled on the performance assessment of foreign trade and based solely on tangible and countable items such as hotel beds, food products, and purchased goods. The bottom line of the Directorate's report was that compared to Bulgaria's foreign trade the earnings of tourism cost only one-quarter of the expenditure on goods production for export, which led it to conclude that Bulgaria's international tourism had "great benefits" for the state economy. Yet, the report also underlined the low number of Western visitors, who were a direct source of hard currency, unlike socialist citizens whose bills were settled through bilateral clearing agreements. At this stage, foreign guests in Varna Resort came predominantly from the socialist bloc (nearly 70 per cent from Czechoslovakia alone) while Westerners accounted for merely 2 per cent of the total number of visitors and an even lower share of revenue, 0.3 per cent.[14]

Once the financial profitability of international tourism was confirmed by the data, resort development was extended to previously non-urbanised coastal areas, and the large-scale construction of entire holiday resorts from scratch was soon as much an investment priority as that of new industrial combines.[15] By the end of the 1950s, Sunny Beach and Golden Sands – the resorts that would become the face of the Bulgarian tourism brand – opened for business. Their speedy completion and modern architecture was acclaimed in party plenums, and the modern standard they epitomised was replicated in smaller holiday developments along the Black Sea coast.[16] Results were swift. In 1957 Bulgaria had attracted 16,776 foreign holiday-makers, all of them accommodated in Varna Resort; in 1959 their number surged to 62,200, and two years later it rose again to 120,000.[17] Tasked with increasing its financial gains exponentially, Balkantourist sought to attract more Western tourists, and formed a separate management section for "Capitalist countries."[18]

To monitor its profit generation from foreign tourists, the government regularly appointed inter-ministerial commissions with representation from the commercial, financial, and banking branches of the state bureaucracy. The complex nature of Bulgaria's tourism finances compared to manufacturing and trade led to disagreements over the proper method of measuring tourism profits. However, even though the audited organisations found the accounting models rather conservative, the numbers confirmed that tourism yielded higher returns than exports did.[19] In fact, tourism was the only source of invisible earnings that promised to offset the persistent trade deficit with the West.

International tourism was bound to foreign trade not only strategically but also institutionally. As Balkantourist still lacked a network of travel bureaus abroad, it relied on foreign-trade representation to broker its offers to foreign agencies.[20] Yet, from the outset Balkantourist's management complained that this style of doing business was inadequate: Bulgarian trade envoys lacked specialised expertise to promote tourism in Bulgaria, and moreover seemed less interested in securing tourism deals, which were an addendum to their annual plans for commodity trade.[21] Arguing for the urgency of establishing its own bureaus abroad, Petar Ignatov, Chief Director of Balkantourist, explained in a report to his supervising minister:

> The nature of the tourist trade – unlike the other [trade in goods] which works mainly by way of one-off deals – requires constant contact with both the travel bureaus that attract tourists and the tourists themselves, combined above all with unremitting attention from the moment services were offered and purchased until the moment when [the tourists] left our country.[22]

Yet, it was to the advantage of tourism to be coupled to foreign trade, and while lobbying for more operational autonomy, Ignatov also did not miss the chance to push tourism in trade talks, even advocating that tourists be added to the annual commodity lists in bilateral trade protocols.[23]

Although without the institutional network of foreign trade, tourism was quite successful in setting its own rules of commerce. From early on, it managed to circumvent official exchange rates by securing special "premiums" on tourists' hard currency, and by strengthening its purchasing power with "service discounts."[24] Border regulations were another area where Balkantourist vied to legalise tourist exemptions, with the result that visa formalities were increasingly liberalised from 1959 onwards, often in disregard of mounting protests from the powerful security apparatus.[25] Transportation, and especially air traffic, was crucial to the growth of tourism, and Balkantourist succeeded in negotiating its own preferential tariffs with the national aviation authorities.[26] All in all, without having had much standing in the state bureaucracy, in just a few years Balkantourist had managed to shed its inherent structural and ideological restraints in order to contend for its share of international mass tourism.

The entanglement of international tourism and foreign trade: economic and institutional growth in the 1960s

When negotiating with higher-ranking administrative bodies for the further easing of travel regulations, Ignatov regularly pointed to the upsurge in tourism globally, arguing that Bulgaria's participation in this dynamic market would secure much-needed hard currency. Such forecasts were invariably supported by comparisons of some sort, often quite crude, of the profitability of foreign trade and tourism; fundamentally reductionist for both trade and tourism, they revealed a good deal about the theoretical difficulty of embedding commercial tourism into a socialist economy, and the pragmatic challenges of organising its business. In a planned economy, with its materialist mindset and contempt for services as unproductive labour, it was goods that mattered, and calculations of profitability usually boiled down to production costs and hard-currency returns.[27] So while government commissions compared occupancy rates in holiday facilities to sales of goods, the director of Balkantourist did his best to inflate the beneficial ratio of tourism to foreign trade based on the low share of goods (25 per cent) in the overall expenditure in tourism, disregarding capital investment and service provision as seemingly cost-free intangibles.[28]

While arguing that the product offered was akin to commodity exports in terms of its contribution to the economy, Balkantourist's management was well aware that marketing tourism was nothing like trading goods, at least not from the viewpoint of consumption. Emphasising the "service" nature of tourism might not have been the right angle to safeguard its better position in the socialist economy, yet service was indeed what made the tourist's experience satisfactory or not. Ultimately, service was the yardstick by which the customer evaluated the products that Balkantourist offered. Raising the quality of this "product" – that is, of tourist services – was in many ways beyond the control of the service provider alone. It was dependent on the quality of production and supply in a myriad of other sectors: architecture and construction for building hotels modern enough for Western tastes; urban planning and sanitation for maintaining proper standards of cleanness, water supply, and sewerage during the seasonal spike in the summer; transportation for bringing people in and helping them move around the country;

heavy and light industry for the supply of furniture and appliances, foodstuffs and souvenirs, cosmetics and fabrics, beyond the mass production for the domestic market and above its level; labour and education policies for the stable employment of adequately trained personnel in hotels, restaurants and tourist shops. None of these crucial conditions was easy to come by, and Bulgaria's tourism management engaged in painstaking negotiations with the respective ministries to bridge the gap between what was customary for the local population and what the foreign tourist expected to find.[29]

Unlike exported goods that reached Western customers in the familiar setting of their local supermarket, in tourism it was the other way round, as consumers were brought to the product, thereby exiting their comfort zone. The tourist "product" could not be fully detached from the broader environment of socialist Eastern Europe, not even behind the gates of the resort complex. Illustrating this dilemma, in 1959, a Swiss tourist sent a lengthy account detailing all the deficiencies in service she had encountered during a holiday trip around the country. Thickly underlined by the administrator who reviewed her letter, she concluded that bad service made tourists "suddenly feel as if in an alien environment," and further explained the profound difference between cultural exoticism and cultural shock in a strikingly colonial language: "The tourist has no trouble accepting the customs of the visited country no matter how different from his own. For example, he will not be offended to see in front of the house a Negro who gulps gruel with his bare hands. But in any country he goes, even in Africa, the tourist wants, in the hotel as well as in the restaurant, to find his own habits."[30] While recognising the overwhelming problems with the services they provided, Balkantourist lacked the institutional weight to tackle them across the board.

In 1963, a new government body, the Chief Directorate for Tourism (CDT), was established under the Council of Ministers. Its primary mission was to coordinate all economic activities in the field of tourism with those in other sectors of the economy, and to actively pursue international cooperation. Balkantourist – along with other organisations in charge of tourist accommodation, restaurant catering, and specialised travel – was subsequently switched from the MFT to the CDT as an enterprise specialised in international tourism. Shortly after the reshuffle, Balkantourist was finally authorised to set up its own network of tourist bureaus abroad.[31] In 1965, Balkantourist's prerogatives expanded further when it was rebranded as an Incorporated Trade Enterprise for International Tourism, with its own budget. In addition to its previous activities, it could now operate bureaux de changes in resorts, stations, airports, international hotels, and at border checkpoints; organise day trips for foreign guests abroad (with destinations such as Istanbul, Odessa, and even Cairo); arrange re-export and compensatory deals; help establish foreign travel bureaus in Bulgaria; and set up joint ventures specialising in tourism in the West.[32]

From the very beginning Balkantourist used a concept of tourism that implicitly was limited to the profit-generating category of holidaymakers, but the definition was at the same time broad enough to include any type of travel, regardless of its actual purpose. Transit mobility, which generated income from visa fees as well as spending on the road, thus accounted for a substantial share of Bulgaria's international tourism, adding hundreds of thousands of visitors to the statistics along with hard-currency revenues. Though geographically distant from Western tourist markets, Bulgaria benefited as a crossing point for seasonal workers from Turkey and the Middle East travelling to Western Europe and back. More than a ruse to boost tourist numbers, transit travellers were perceived as a crucial segment that required different services and infrastructure investment beyond the resort complexes.[33] In 1965, when 1 million foreign tourists visited the Bulgarian "vacationland," as RFE reported it, in fact only 98,593 out of 634,756 Western "tourists" were in Bulgaria on holiday. The rest were all transit passengers, most of whom were Turkish citizens.[34] Parallel statistics based on hotel occupancy showed that, despite specifically targeting foreigners, "economic tourism" in reality catered predominantly to Bulgarian nationals.[35] As the operational concept of "economic tourism" was quite elastic, defined by financial receipts and not by purpose of travel, the data on Bulgarian "tourists" (aggregated by hotel check-ins) was inflated by all sorts of official trips, just like the number of "foreign tourists" (counted by border entries) was significantly boosted by transit passengers.[36] Yet, as the annual plans for the fledgling tourism sector prioritised the increase of hard-currency receipts, the reform of "economic tourism" remained firmly dedicated to making Bulgaria more attractive to Western visitors.

At the Ninth Party Congress in November 1966, the BCP leadership made much of tourism as a highly efficient sector of the national economy.[37] A month later, the CDT was replaced by the CT, which was given the rank of a ministry in the next round of administrative reform in 1973.[38] Even before the CT acquired a ministerial status, its functions had greatly expanded, intersecting, often competitively, with many other economic areas – large-scale construction, transportation, light industry and retail, various supplementary services, and, as soon as specialised tourism colleges were founded, even higher education. The supervisory role of the CT was also boosted by being put in charge of all tourism-related activities run by ministries, city councils, and economic units.[39] Moreover, capital investment in international tourism was legally qualified as a "national priority," and thus elevated to the same level of importance as central government buildings and major socialist monuments.[40]

Resort developments soon expanded beyond the Black Sea coast. Mountain resorts with winter sports facilities such as Aleko-Vitosha, Pamporovo, and Borovets extended the calendar for international tourism. Various forms of specialised tourism were also on offer: congresses, balneological treatments, hunting expeditions, and weekend breaks. By then, it was a given for the tourism administration that the resort boom was driven by an influx of foreigners, and responsibility for domestic "social tourism" was largely relegated to employers. At the national conference to mark the International Tourist Year in 1967, a high-ranking tourism official hailed the country's recreational base as servicing exclusively foreigners, and the role of the Bulgarian worker as producer rather than consumer:

What the imagination of Homer gave birth to in the Bronze Age [*sic*] – his fascinating protagonist Hephaestus building with his magic hammer palaces of gold on the slopes of Olympus – today in our country, our people actually created along the Black Sea coast, on the slopes of the Rila and Rhodope Mountains, an entire necklace of golden palaces, and delivered them for the needs of international tourism.[41]

By 1969, international tourism brought in hard-currency revenues amounting to 5 per cent of trade exports, and was thought would grow to 20 per cent in the next five years.[42] At the time, Bulgarian exports were still primarily structured around agricultural products and processed fruit and vegetables, which faced rising trade barriers in their main markets in Western Europe due to the consolidation of the customs union of the European Economic Community (EEC). While a serious impediment to Bulgaria's foreign trade, it was also an opportunity to further highlight the economic importance of tourism. Instrumental comparisons of profitability vis-à-vis foreign trade focused even more directly on the circulation of goods. In its report for 1970, the CT included a lengthy appendix to demonstrate that basic foodstuffs (fresh and canned fruit and vegetables, meat, alcoholic beverages, etc.) yielded three- to tenfold higher profits in hard currency when sold domestically to foreign tourists than when exported to Western buyers by the foreign-trade organisations.[43]

In the early 1970s, international tourism, framed as "export in situ," took on new significance as a segment of the national economy following new EEC restrictions on bilateral trade between member states and the Eastern bloc. In 1970, the CT submitted its first comprehensive strategy for the development of the tourist sector over the next five-year period. Reporting on the recent surge of the tourist industry worldwide, not just in Bulgaria, CT Chairman Petko Todorov requested an accelerated expansion of the material base of international tourism, because according to his estimates the demand for Bulgarian holidays in capitalist markets already exceeded availability. For the five years of its existence, the CT had witnessed a surge of 219 per cent in economic tourism measured by hotel occupancy – from 8,909,000 nights in 1965 to 19,544,000 in 1969 – outpacing dramatically the growth of social tourism (from 10,190,000 to 12,506,000 nights). While Bulgarians still dominated "economic tourism" (thanks to short leisure trips, cheaper tourism options in the country's interior, and domestic business trips by public officials, artists, scholars and so on), in the so-called "seasonal base" – meaning the facilities solely for holiday use – the share of foreign customers was a steady 80 per cent throughout the later 1960s, when data collection allowed for such disaggregation.[44] Due to its recognised contribution to the national budget, the CT was now authorised to use 2 per cent of its own hard-currency receipts (in addition to 1 million convertible lev of export revenues from the MFT) to import consumer goods to meet foreign tourists' needs.[45]

Over and above the statistics that were mobilised specifically to demonstrate the accomplishments of the tourist sector, however, the documents reveal a parallel success story, perhaps even more impressive albeit often overlooked, of institution-building. Since its foundation, the CT had managed to intervene in numerous long-established sectors of Bulgaria's planned economy, setting entirely different production standards, priority quotas, and supply lines solely to advance its international operations. From carpentry and faience manufacturing all the way to the construction of highways and dams, international tourism set its own agenda in the national economic plan.

Breaking ground in the Scandinavian market: the cooperation of foreign trade and international tourism

From the late 1950s onwards, as part of its opening up to the West, the Bulgarian state intensified both its economic and its political contacts with Scandinavia, a region that it had had little interaction with in the formative years of state socialism and prior to the Second World War. As these efforts coincided with the pioneering steps to develop international tourism at home, it naturally played a part in Bulgaria's advances to the Scandinavian countries. While the Bulgarian diplomatic corps saw Sweden as the stepping stone in the region, the tourist agencies thought Denmark would be their entry point to the Scandinavian market. At the time Bulgaria only had a diplomatic mission in Stockholm, though in 1957 the resident ambassador was also accredited to Denmark and Norway.[46] A year later, however, Bulgaria signed an Air Transport Agreement with Denmark, and Copenhagen became one of the few Western European airports in the Bulgarian civil aviation network. The agreement also led to the establishment of an office of the Bulgarian airlines (later named Balkan) in Copenhagen, around the time when Bulgaria also opened a trade mission there under a bilateral trade agreement signed in 1959.[47]

Though political contacts between the Scandinavian countries and Bulgaria were minimal, the legation in Stockholm still played a role in advancing the national objectives of economic cooperation. Tourism was a particularly suitable niche for the ambassador's broad diplomatic mandate, because the large travel companies covered the entire Scandinavian region. In 1959, the legation informed the Ministry of Foreign Affairs (MFA) of the Swedes' and Danes' growing interest in holiday destinations in Bulgaria, and particularly the new Varna Resort.

The embassy in Stockholm had been approached by the chief executive of the Scandinavian subsidiary of Wagons-Lits Cook, and Balkantourist did not fail to note such a promising business opportunity. It highlighted Scandinavia alongside traditional tourist providers such as Austria and West Germany as a market in need of more concerted exploitation.[48] Stationing representatives in the West independently of the trade missions was still wishful thinking, but Balkantourist sought to forge personal ties with foreign travel agencies through business trips to the region. In 1959, Petar Ignatov proposed opening a handful of foreign bureaus, including one in Sweden, while simultaneously sending business delegations to eight Western countries, including all the Nordic countries barring Iceland.[49] The green light for such initiatives depended on the MFT, which was reluctant to sideline its own envoys. Ignatov's bold demands were rarely fulfilled, but a compromise was usually found, which was why a year later he was given official permission to invite foreign travel bureau representatives to Sofia to gain first-hand experience of Bulgarian resorts and Balkantourist's business style. Danish company representatives were among Balkantourist's first Western guests.[50]

Barred from setting up its own branches abroad, Balkantourist began to collaborate with the Bulgarian civil aviation authorities, which already had a small international network. In 1960, the two organisations put together several all-inclusive tours for fortnight-long stays at Bulgarian seaside resorts. These offers were only marketed in three or four Western European countries, but Copenhagen was a key hub from the outset. The Copenhagen–Varna package was promoted across the entire Nordic market by soliciting options for regional transfers with Scandinavian aircraft from the other three capitals plus Bergen, Gothenburg, Kristiansand, and Stavanger.[51] At the time, the flight time from Copenhagen to Sofia alone was over 13 hours, followed by a domestic flight to the coast a day later. This made a Black Sea holiday quite a feat for Scandinavian holidaymakers, especially if they lived outside the Danish capital, and in fact the first Danish tourists registered by the MFA date from 1962.[52]

Meanwhile, commerce between Denmark and Bulgaria was growing. Once the MFT had set up shop in Copenhagen in 1960, bilateral trade between the two countries increased in the first two years from $0.6 million to $1.2 million. By 1965 it had grown to $3.8 million, with Bulgaria's trade deficit skyrocketing from $216,000 in 1960 to over $2 million in 1965.[53] The country's weak exports could not pay off the costs of high-tech imports of factory machinery and equipment from Denmark, but tourism represented a potential remedy.

Acknowledging that Bulgarian airlines' small aircraft and infrequent flights did not add to the attractiveness of all-inclusive tours from Copenhagen, Balkantourist looked for new partners in the Scandinavian market. This strategy became possible once tourism had been hived off from the foreign-trade sector (with the establishment of CDT in 1963). In 1964, Balkantourist signed its first contracts with two Danish travel companies with an all-Scandinavian reach, Startour and Jørgensens. In 1965–1966 alone, the number of Danish package tourists in Bulgaria doubled (from 1,219 to 2,533) and the revenues from this tourist flow almost tripled. A year later, the two Danish companies' package holidays took 4,503 Danes to the Black Sea coast, with a proportionate increase in revenue.[54] At a point when the Bulgarian resorts still counted Western holidaymakers in the tens of thousands, Scandinavian visitors had a visible presence. Moreover, the Bulgarian tourist management noted that the Scandinavians, who enjoyed high living standards, generally spent more money on their holidays. In fact, a few years later, Swedish tourists were estimated to bring in the highest hard-currency revenues per tourist.[55]

Under the management of CDT, Balkantourist was finally authorised to set up its own network of foreign bureaus in the mid-1960s, and it began with 16, nine of them in Western Europe, including both Sweden and Denmark.[56] Balkantourist's ambitions did not end there, and soon the enterprise was vying for bilateral tourism agreements with the UK, Benelux, and Scandinavia where "our interests are big."[57] Thus far, economic cooperation with Denmark had run ahead of active intergovernmental relations, and economic ties were effected by a few business-minded people in Bulgaria's trade missions abroad and the small-scale tourist enterprise at home. A breakthrough in the diplomatic stalemate between Bulgaria and Denmark came in 1967 with an exchange of visits between the foreign ministers, Ivan Bashev and Jens Otto Krag.[58] As a result, the two governments signed several new agreements testifying to their mutual desire for greater cooperation in various fields including trade and tourism. The Agreement on Economic, Industrial and Technological Cooperation laid the ground for the establishment of a bilateral agency, the Mixed Bulgarian–Danish Commission for Economic, Industrial, and Technical Cooperation (MBDC). The MBDC met annually to discuss proposals for joint industrial projects, educational exchanges for specialists, the transfer of technical know-how, and industrial assistance, and then matched the offers and requests with suitable organisations in their respective countries.[59]

Similarly, the Agreement for the Suppression of Visas aimed to improve Bulgaria's position in the Danish tourist market.[60] Though this relaxation worked both ways, it had less impact on the mobility of Bulgarian citizens, who still faced formidable obstacles in exiting their own country. Indeed, the flow of travellers between the two countries remained disproportionate, so that in 1972, for example, the number of Danish visitors to Bulgaria was twenty times greater than the number of Bulgarians travelling in the opposite direction.[61]

Tourism had been on the Bulgarian MFA's agenda for the talks with Krag, although less prominently so than trade. The Danish government, however, had its own motives for raising the issue, inquiring about the possibilities for private Scandinavian companies to develop their own resorts on the Bulgarian coast. Bulgaria had already made legal provision for joint ventures with foreign capital (in trade as well as tourism), but so far such companies were only permitted on foreign soil, and with majority holding rights remaining in Bulgarian hands. Allowing foreign ventures to operate in Bulgaria was an entirely different matter, and highly sensitive politically because it challenged the fundamental socialist principle of state ownership. The MFA nevertheless committed itself to taking this question in consideration, and seemed to be positively inclined.[62]

As a result of the intergovernmental talks, the importance of economic cooperation with Denmark was reaffirmed and the plan targets were significantly increased. For the CT, this meant a projected threefold increase in Danish tourists in the coming three years (1968–1970).[63] However, this was the very period when tourist numbers from Denmark plummeted. The foreign representatives cautiously attributed this to the devaluation of the Danish krone and increased taxes, which had hit the tourist market hard and sent some travel agencies into bankruptcy. Among them was Jørgensens, which had accounted for the greatest number of Danish tourists in Bulgaria.[64]

The invasion of Czechoslovakia by the Warsaw Pact in 1968 was deliberately downplayed by the reports, although its impact was arguably more significant. In fact, Startour cancelled its tours to the country indefinitely after the invasion. Just as international politics could boost tourism, the opposite proved also to be true. Mirroring the Danish government's decision to rescind its invitation to the Bulgarian leader Todor Zhivkov to visit in September 1968, Startour publicised the cancellation of its tours in the Danish media as a political rather than a business decision.[65] Instead of tripling as planned, by 1969 the number of Danish tourists in Bulgaria had dropped to one-third of the 1967 level. Yet even in such a poor year, tourism still kept its weight in the balance of payments. In 1969, the CT reported its hard-currency income by country relative to export revenues from bilateral trade: revenues from Danish tourists were estimated to make up 9 per cent of Bulgarian exports to the Danish market, while those from Swedish tourists amounted to 26 per cent of Bulgarian exports there.[66]

There was no denying that 1968 was disastrous for international tourism in Bulgaria. The previous year, the government had temporarily lifted visas to mark the International Tourist Year, and this special provision had then been extended to 1968. Nevertheless, a drop of 12 per cent in the number of Western tourists and 17 per cent in tourist revenues was registered by the CT.[67] As the repercussions from the invasion of Czechoslovakia began to hit, the BCP mobilised the diplomatic corps to improve Bulgaria's image. In May 1969, the Politburo passed a resolution calling for greater liaison with the Scandinavian states, and Bulgaria finally opened an embassy in Copenhagen. With the assistance of the diplomatic corps, the CT managed to sign new contracts with two Danish travel companies (Spies and Danropa) and to restart its work with Startour. The new partnerships secured the return of 4,600 Danish holidaymakers to Bulgarian resorts in 1970.[68]

Bulgarian–Danish tourist partnerships in the 1970s: opportunities and failures

In the 1960s, Bulgaria had mainly targeted Danish tourists through the travel programmes of Scandinavian tour operators. In the 1970s, the CT began to pursue a more active role in the management of the tourist groups. In the 1960s, bilateral partnerships had been negotiated by the small team of four representatives of Balkan Airlines and CT stationed in Copenhagen.[69] In contrast, in the 1970s new actors on the Bulgarian side became involved in brokering international tourism, while direct personal contact with foreign firms was replaced by more complex partnerships that required the coordination of multiple bureaucratic apparatuses.

One actor joining the promotion of Bulgarian resorts in Denmark was the newly opened Bulgarian embassy in Copenhagen. Working with the trade envoy, the chargé d'affaires launched a number of initiatives to diversify their tourist partnerships and reach out beyond the large travel companies, which were seen as monopolising the market and undercutting their Bulgarian partners.[70] One of the embassy's first successes was with Folketurist, a travel bureau that had close ties to the Danish Communist Party. Folketurist mainly organised holidays to Eastern Europe, "above all to popularize the socialist countries" as their director said during his meeting at the embassy. Folketurist had previously sent 500–600 people in groups to Bulgaria by selling Startour and Jørgensens package tours, but Spies, Bulgaria's main partner in Denmark, refused to take additional groups contracted by external travel agencies. Folketurist expressed interest in expanding in Bulgaria by chartering its own flights, having already established similar programmes in the German Democratic Republic (GDR), the USSR, and Romania, where it sent several thousand people a year. Certainly, the partnership was an attractive prospect for Bulgaria. Folketurist's interest in specialising in low-season holidays for trade unionists, young people, and party activists fitted well with Bulgaria's attempts to stagger the holiday rush. The broader interest in exploring the country, encouraged by Folketurist's semi-political profile, also chimed with the new Bulgarian strategy for diversifying its international tourism.

The only problems that the Bulgarian embassy encountered in its preliminary negotiations with Folketurist were the company's competition with Spies and its complaints about the high prices of the Bulgarian airlines compared to the Scandinavian carriers. In his efforts to convince Folketurist's management to concentrate on Bulgaria, the Bulgarian chargé d'affaires found an unexpected ally in the Bulgarian Central Committee of Anti-Fascist Fighters, which, coincidentally, had invited Folketurist's director to a meeting in one of Bulgaria's seaside resorts.[71]

Political expediency and the bonds of socialist solidarity could advance as much as injure business interests. In their efforts to increase profits, the Bulgarian tourist agencies adopted a business model that had less to do with the socialist credo than with capitalist entrepreneurship. Yet, there were limits to the official tolerance of activities at variance with their ideological commitments. The collaboration with Swedish Folkturist illustrated this, for once it had failed to meet its financial obligations to Balkantourist and Balkan Airlines and the two Bulgarian agencies threatened to take legal action, the Central Committee of the Swedish Communist Party appealed to the Bulgarian embassy in Stockholm to prevent the company's bankruptcy. Under pressure from the MFA, the CT withdrew its ultimatum and agreed to reschedule Folkturist's debt, against its own best interests.[72]

Apart from such overlaps – and contradictions – between economic and political interests, the attempt by various arms of the Bulgarian bureaucracy to work together often suffered from a lack of coordination, and business deals were more likely to be undermined by incompetence than by political reasoning. On many an occasion, the involvement of institutions at home jeopardised what Bulgaria's foreign representatives achieved, bringing a certain dissonance to Bulgarian–Danish negotiations. While the CT officials abroad concentrated on actual deals, and saw them through from beginning to end, their superiors would often fly in to sign the contract without taking the time to study the specific conditions or the nature of recent business relations. In 1972, for example, the CT delegation for the annual renewal of the tourist contract with Spies, the largest in Denmark, obtained "deplorable results" in the words of the local representative. By restarting negotiations from scratch instead of simply signing the agreement that had already been approved, the delegation seriously disrupted the smooth operation of this long-term deal. Moreover, during its visits to Copenhagen, the delegation did not even keep the embassy and the trade mission in the loop.[73] This led to the Danish company not only rejecting the new demands, but cancelling all the groups that were already confirmed for the coming year.

Upheavals because of poor coordination were most frequent in the work of the MBDC. Under the supervision of two government bodies – the Danish Committee on Industrial–Technological Cooperation and the Bulgarian State Committee for Science and Technical Progress – the MBDC was expected to provide a stable framework for bilateral economic relations by facilitating partnerships between Bulgarian state enterprises and Danish private companies. The MBDC's objectives gave it considerable scope; however, it soon became clear that the national delegations, having different economic goals, also had very different priorities for the MBDC. This clash seemed insurmountable, blocking any real way forward, which led the chairman of the Danish party to threaten a boycott of the annual sessions. The crux of the problem was that the Bulgarian side avoided committing itself on trade (largely because it dared not interfere with foreign trade's parallel chain of command) and focused instead on industrial assistance. In contrast to the passivity of the Bulgarians, the Danes, who were not only state officials but also private entrepreneurs, were keenly interested in finalising deals that could guarantee a financial return.[74]

The Bulgarian side of the MBDC was tasked with increasing machinery exports, and tourism was not initially on its agenda. This perhaps explains one of the Bulgarians' first serious gaffes. At the second session of the MBDC in 1969, the Bulgarian delegation was informed of ongoing negotiations between the Bulgarian airline company and a Danish architect for a seaside hotel. Recognising the potential of this contact to grow into a larger economic cooperation venture, both parties to the MBDC agreed to set up a special commission with broader institutional participation, including the Bulgarian ambassador and Danish MBDC delegates, and possibly enlisting the help of Scandinavian Airlines System (SAS).[75] By the end of the following year, the Danes had fulfilled their commitment, but the Bulgarians remained silent and even failed to pay the architect for his work. At this point, the head of the Danish party, Knud Hannover, brought the issue to the attention of the Bulgarian embassy in Copenhagen.[76] In response to the latter's diplomatic memo on the issue, the Bulgarian MBDC party informed the MFA that

> there is no trace in the Bulgarian part of the Commission of the matter that the memo discusses, it is neither known which is the Danish firm in question, nor which Danish representatives have come to Bulgaria in this regard, nor with whom they have held negotiations. The efforts made to learn something about this issue did not yield any results.[77]

The issue kept coming up for at least a year, with no progress on the Bulgarian side to resolve it, to the growing irritation of their Danish counterparts.[78] That was the tipping point for the Danish Chairman, and soon after he threatened to dismantle the MBDC.

Ideological volatility of Bulgaria's international tourism

Many of the problems with the organisation of Danish tourism in Bulgaria were caused by poor coordination among the institutions mired in the bureaucratism of its socialist command economy. However, tourism behind the Iron Curtain also faced problems that were quite mundane in essence, but took on an ideological form that sparked political reactions and sometimes even public panic. Such incidents were at times caused by random, unforeseeable factors, at other times by business disputes, but in either case their unexpected escalation into an ideological clash highlighted the possibility, real or imagined, of Western holidaymakers in Eastern Europe caught in the crossfire of Cold War antagonism. On some occasions, the Western press was quick to resort to ideological clichés, inciting public fears of the authoritarian environment of socialist Eastern Europe where Western tourists could find themselves vulnerable to state harassment and repression. The Bulgarian institutions, for their part, were overly sensitive to what they viewed as "hostile propaganda," and often suspected political orchestration when there was none.

More than any other form of cooperation across the Iron Curtain, East–West tourism was influenced by the conflicting impulses of the Cold War and detente. It brought people from both sides of the divide together, and its role as emissary in securing a peaceful coexistence was a frequent trope in the rhetoric of tourism. In 1966, for example, Petko Todorov, Chairman of the CDT, opened his report to Todor Zhivkov with the usual reference to the global surge of international tourism, which led him to the somewhat bizarre conclusion that "In one year alone, in the orbit of international tourism, [we see] more people taking part, of their own free will and with best intentions, than the number of those involved in the entire Second World War."[79]

While Bulgaria's trade and tourist partnerships in the Danish market in many ways developed in parallel, complementing one another, there was one significant difference in their operation, which became painfully clear after the events of 1968. Although both types of business deals were negotiated across the Iron Curtain between like-minded professionals whose pragmatic interests superseded ideological disagreement, tourist contracts were far more volatile, ideologically speaking. Unlike exports of consumer goods, a tourist product tailored to the Western client and sold to the Danish market had to be consumed "behind enemy lines." Bulgarian resorts were designed to shelter foreign guests from their surroundings, but any rise in geopolitical tension impacted on the tourists' sense of safety. This was also the case with far more trivial holiday disruptions such as flight delays or road accidents.

In 1969, when Western tourists still had the Warsaw Pact invasion of Czechoslovakia fresh in their minds, a road accident caused by a Danish tourist coach received a great deal of attention in both the Danish media and Bulgarian diplomatic correspondence. The Danish driver, who was the owner of the travel bureau that had arranged the excursion, hit a little girl on the road (causing injuries leading to concussion) and was subsequently detained while the accident was investigated.[80] The 24-day arrest of a Danish citizen was covered extensively in the Danish press, where his personal account of police harassment and interrogation under torture became the main story.[81] This took on political overtones because of the driver's insistence that the Bulgarian prosecution suspected a political conspiracy was behind the accident, and that his interrogators had tried to force him to admit a political motivation. The Danish side of the story thus echoed Cold War spy-novel tropes of authoritarian harassment, while for the Bulgarian diplomatic service the "hostile press campaign" smacked of a propaganda operation.[82]

This two-sided ideological rhetoric completely overshadowed the more prosaic aspects of the story: the concerns of a Danish small businessman at the financial loss incurred by his arrest, and the fears of a country desperate to attract Western tourists about its effect on prospective customers. In the same turbulent year, the CT affiliate in Stockholm reported on another incident that threatened to hurt its work in the Swedish market: the news of an outbreak of jaundice at Sunny Beach was circulating among the Swedish travel firms.[83] In a response marked "extremely urgent," the MFA instructed the embassy to immediately put out a denial of what it was to describe as a "malicious rumour."[84]

Ideological rhetoric and pragmatic considerations played a role, sometimes purely opportunistically, in regular business deals in the tourist sector. At the end of the 1960s, the relationship with Startour was one example of this interplay between business strategy and political rhetoric. In 1968, the company announced its refusal to do business with Bulgarians in protest at the invasion of Czechoslovakia. The following year, the Bulgarian embassy reported another cancellation by the company, which had apparently resumed its business in Bulgaria after all. However, the ambassador, who communicated Startour's grievances and tried to save relations with the company, explained that they had cancelled because of the competition from Bulgarian state agencies, which were targeting Danish customers directly by undercutting Startour's prices.[85]

In the years that followed, Scandinavian participation in the Bulgarian tourist sector expanded. In 1975, more than 45,000 Scandinavians visited the country.[86] Business disputes occasionally escalated into public mudslinging, complete with Cold War stereotypes. In 1975, for instance, Tjæreborg Rejser, the largest Danish tour operator with branches throughout Scandinavia, decided to terminate its partnership with Balkan Airlines. At this point Tjæreborg accounted for around 70 per cent of Danish tourists in Bulgaria and for an added contingent of West Germans flown by Balkan (around 5,500 tourists).[87] Tjæreborg found Balkan's timetables inconvenient, and somehow the disagreement spiraled into open conflict. The West German news magazine *Stern* published a lengthy piece on the matter, opening with a swift political judgement: "[This is] how the bureaucrats from the Eastern bloc (socialist countries) spoil the holiday break of German tourists."[88] Bulgaria took its publication as evidence of "an attempt to use this case for political purposes and to blackmail our country."[89]

The Bulgarian authorities saw the ideological card as one played to weaken their position in business deals or to damage the economic interests of the country. However, in the context of detente they also sought to exploit it to their benefit. The new phase of European cooperation heralded by the Helsinki Accords of 1975 promised new opportunities for East–West tourism. While the Bulgarian regime viewed the "third basket" of the Final Act with growing suspicion for its humanitarian focus, the CT welcomed the inclusion of tourism under the rubrics of both economic cooperation and human contacts.[90] The promotion of international tourism featured in both second and third baskets, in recognition of "the interrelationship between the development of tourism and measures taken in other areas of economic activity," and of tourism's contribution to "the growth of understanding among peoples, to the improvement of contacts and to the broader use of leisure."[91] For the CT, this recognition encouraged a new approach to attract Western visitors, one relying on diplomatic channels far more than on commercial partnerships, and on intergovernmental agreements rather than business deals.

Conclusion

In placing Bulgaria's tourist boom against the Cold War backdrop, RFE's 1966 report quoted earlier in this chapter speculated on whether opening up to the West might soon be halted because of the political risks it posed to an authoritarian regime with anti-capitalist zeal, or whether the financial incentives would prevail over ideological fears. Tackling the political implications of Bulgaria's new economic course, as represented by its Western-oriented tourist industry, the report even suggested that it might be symptomatic of a larger process of "polycentrism, desatellization and the attenuation of the Cold War upon the Eastern European countries."[92] This hypothesis was soon proved wrong when in 1968 the Warsaw Pact tanks rolled into Czechoslovakia and Bulgarian troops were mobilised to join in the invasion. Domestically, the crackdown on the Prague Spring put an abrupt end to tentative reforms of Bulgaria's planned economy, and the country's economic liberalisation was largely curtailed.[93]

However, RFE's more pessimistic scenario – that political expediency would dictate that Bulgaria downscale its tourist services for Westerners – was not warranted either. While its post-1968 domestic policies were no doubt shaped by concern at the political risks that opening up to the West posed to regime stability, international tourism enjoyed the unwavering support of the party apparatus throughout the 1970s. The expansion of recreational services for foreign guests, primarily Westerners, remained an economic priority. While Bulgaria welcomed an ever-increasing number of holidaymakers from non-socialist countries, its political system remained firmly in the orbit of the Soviet Union and there was no intention to pursue any form of "desatellization."

From the early 1960s, socialist Bulgaria pursued an ambitious programme of international tourism that targeted particularly Western visitors who could contribute to the country's hard-currency revenues. The aims behind the tourist boom were closely linked to the country's economic policies, and especially its foreign trade, but the means to that end bound the tourist sector to diplomatic and foreign-policy endeavours. In Bulgaria's tourist partnerships with Denmark, the triangle of trade, tourism, and diplomacy took on different shapes over time. In the early 1960s, tourism assisted foreign trade, and the two together were instrumental in establishing relations with Denmark, thus paving the way for high-level intergovernmental relations towards the end of the decade. In the 1970s, the tourist sector continued to act as a vehicle of national image-making, but, reciprocally, the diplomatic service became a promoter of international tourism. The close links between tourism and diplomacy, however, showed up in less advantageous ways too. Tourism behind the Iron Curtain was easily hampered by any event that could be framed as symptomatic of East–West political divergence.

Notes

1 The research leading to this chapter is part of the project PanEur1970s, which has received funding from the European Research Council (ERC) under the European Union's Horizon 2020 research and innovation programme (Grant Agreement No. 669194).

2 Elitza Stanoeva, "Sotsialisticheskata targoviya v Balgariya (1954–1963): Ideologiya, distsiplina i marketing" [Socialist trade in Bulgaria (1954–1963): Ideology, discipline and marketing], *Sociological Problems*, nos. 3–4 (2015).

3 Report of Petko Todorov, chairman of CT, to the Chairman of the Committee for Economic Coordination regarding the development of "International and internal tourism" during the sixth five-year plan 1971–1975, 8 August 1970, Central State Archives of Bulgaria (hereafter TsDA), f. 1230 (Committee on Tourism), op. 1, a.e. 54, l. 4.

4 For subsidized domestic tourism as a proletarian social right, see Wendy Bracewell, "Adventures in the Marketplace: Yugoslav Travel Writing and Tourism in the 1950s – 1960s," in *Turizm: The Russian and East European Tourism under Capitalism and Socialism*, ed. Anne E. Gorsuch and Diane P. Koenker (Ithaca: Cornell University Press, 2006), 251; Igor Duda, "Workers into Tourists: Entitlements, Desires, and the Realities of Social Tourism under Yugoslav Socialism," in *Yugoslavia's Sunny Side: A History of Tourism in Socialism*, ed. Hannes Grandits and Karin Taylor (Budapest: Central European University Press, 2010); Duncan Light, "'A Medium of Revolutionary Propaganda': The State and Tourism Policy in the Romanian People's Republic, 1947–1965," *Journal of Tourism History* 5, no. 2 (2013), 189–94.

5 Another form of "export in situ" developed across the Eastern bloc was the hard-currency shop. For the history of the Bulgarian hard-currency shop, Corecom, see Rossitza Guentcheva, "Mobile Objects: Corecom and the Selling of Western Goods in Socialist Bulgaria," *Balkan Studies* 45, no. 1 (2009).

6 For an overview of its economic performance, see Frank W. Carter, "Bulgaria," in *Tourism and Economic Development in Eastern Europe and the Soviet Union*, ed. Derek R. Hall (London: Belhaven Press, 1991).

7 John Walton, "Preface: Some Contexts for Yugoslav Tourism History," in *Yugoslavia's Sunny Side: A History of Tourism in Socialism*, ed. Hannes Grandits and Karin Taylor (Budapest: Central European University Press, 2010), xi–xii; Shawn Salmon, "Marketing Socialism: Inturist in the Late 1950s and Early 1960s," in *Turizm: The Russian and East European Tourism under Capitalism and Socialism*, ed. Anne E. Gorsuch and Diane P. Koenker (Ithaca: Cornell University Press, 2006), 186–204.

8 Vacationland Bulgaria: The Tourist Boom, J. V. Storojev, 24 February 1966, Vera and Donald Blinken Open Society Archive, digital repository of RFE/RL Background Reports (hereafter HU OSA), 300-8-3-15923, p. 1.

9 Report of Petar Ignatov, chief director of Balkantourist, to the Minister of Trade, 1959, TsDA, f. 310 (State Economic Enterprise 'Balkantourism'), op. 2, a.e. 262, l. 44.

10 Historical Information on Balkantourist, TsDA, f. 310, archival inventory, p. 2.

11 Report of Nikola Yotov, chief director of Balkantourist, to the Minister of Trade, 22 August 1958, TsDA, f. 310, op. 2, a.e. 286, l. 32.

12 Report of Petar Ignatov, chief director of Balkantourist, to the Minister of Trade regarding action measures for developing international and domestic tourism in Bulgaria, undated, TsDA, f. 310, op. 2, a.e. 262, l. 34, 38. For the Yugoslav model of foreign tourism, see Igor Tchoukarine, "The Yugoslav Road to International Tourism:

Opening, Decentralization, and Propaganda in the Early 1950s," in *Yugoslavia's Sunny Side: A History of Tourism in Socialism*, ed. Hannes Grandits and Karin Taylor (Budapest: Central European University Press, 2010); Igor Tchoukarine, "Yugoslavia's Open-Door Policy and Global Tourism in the 1950s and 1960s," *East European Politics & Societies* 29, no. 1 (2015). For similar ambitions in neighbouring Romania, see Adelina Stefan's chapter in this volume.

13 Report of Evgeny Mateev, chairman of the Central Statistical Directorate, regarding the operation of Balkantourist in the resort of Varna in 1956 and 1957, 1958, TsDA, f. 310, op. 2, a.e. 286, l. 2.

14 Ibid., 3–3a.

15 Borislav Georgiev, "Arhitekturno-gradoustroystveniya konkurs za nov chernomorski kompleks kray s. Primorsko [Architectural and urban planning competition for a new Black Sea resort complex near the village of Primorsko]," *Arhitektura* 5 (1960), 26.

16 Resolutions of the Plenum of Central Committee of the BCP on 26 October 1959, TsDA, f. 1B (Central Committee of Bulgarian Communist Party), op. 6, a.e. 4034, l. 17 – ᵛ.

17 Report of Evgeny Mateev, chairman of the Central Statistical Directorate, regarding the operation of Balkantourist in the resort of Varna in 1956 and 1957, 1958, TsDA, f. 310, op. 2, a.e. 286, l. 3a; Historical Information on Balkantourist, TsDA, f. 310, archival inventory, p. 3.

18 Information on the work of Section "Capitalist Countries" in the first quarter of 1960, undated, TsDA, f. 310, op. 2, a.e. 287, l. 81.

19 Information regarding the determined returns of tourism according to the report of the Commission of the Ministry of Trade, undated, f. 310, op. 2, a.e. 287, l. 117–19.

20 Information on the work of Section "Capitalist Countries" in the first quarter of 1960, undated, TsDA, f. 310, op. 2, a.e. 287, l. 81; Information on the work and tasks of our trade representations abroad, 29 January 1960, TsDA, f. 310, op. 2, a.e. 287, l. 106–7.

21 Information on the work and tasks of our trade representations abroad, 29 January 1960, TsDA, f. 310, op. 2, a.e. 287, l. 106.

22 Report of Petar Ignatov, chief director of Balkantourist, to the Minister of Trade regarding action measures for developing international and domestic tourism in Bulgaria, undated, TsDA, f. 310, op. 2, a.e. 262, l. 36.

23 Ibid., 39.

24 Ibid., 35.

25 Circular letter for visa issuance and order of stay of foreign citizens in Bulgaria, 19 October 1960, TsDA, f. 310, op. 2, a.e. 287, l. 3–6. For the intervention of the state security in international tourism in Romania, see Dragos Petrescu, "Closely Watched Tourism: The Securitate as Warden of Transnational Encounters, 1967–9," *Journal of Contemporary History* 50, no. 2 (2015).

26 Letter from Balkantourist to the Ministry of Finance, 25 May 1960, TsDA, f. 301, op. 3, a.e. 16, l. 102.

27 Todor Hristov, "The Dry Cleaning of a Socialist Economy: The Bulgarian Debate on the Concept of Service Industry in the Context of the 1967 Economic Reforms," paper presented at Reforming Socialism: Aims and Efforts before and after 1968 (European University Institute, Florence, 25–27 October 2018); Karin Taylor and Hannes Grandits, "Tourism and the Making of Socialist Yugoslavia: An Introduction," in *Yugoslavia's Sunny Side: A History of Tourism in Socialism*, ed. Hannes Grandits and Karin Taylor (Budapest: Central European University Press, 2010), 8.

28 Report of Petar Ignatov, chief director of Balkantourist, to the Minister of Trade, 1959, TsDA, f. 310, op. 2, a.e. 262, l. 44.

29 For similar difficulties with service provision in Soviet international tourism, see Andrei Kozovoi, "The Way to a Man's Heart: How the Soviet Travel Agency 'Sputnik' Struggled to Feed Western Tourists," *Journal of Tourism History* 6, no. 1 (2014): 57–73. For the Bulgarian failure to shape tourist services according to demand, see Derek Hall,

"From 'Bricklaying' to 'Bricolage': Transition and Tourism Development in Central and Eastern Europe," *Tourism Geographies* 10, no. 4 (2008), 417.

30 Letter from G. T. [anonymized by the author for reasons of personal data protection – E.S.] to the Minister [of Foreign Affairs] of Bulgaria, 25 September 1959 (Lausanne), TsDA, f. 310, op. 2, a.e. 263, l. 7 (The original letter is not preserved in the archives, only the Bulgarian translation thereof).

31 Veselin Metodiev and Lachezar Stoyanov, *Balgarskite darzhavni institutsii 1879–1986* [Bulgarian state institutions 1879–1986] (Sofia: D-r Petar Beron, 1987), 71.

32 Statutes for the establishment and activity of Incorporated Trade Enterprise for International Tourism Balkantourist, 23 August 1965, TsDA, f. 310, op. 5, a.e. 2, l. 79–81; Protocol of National convention on questions of tourism organized by the CT, 25 April 1967, TsDA, f. 310, op. 5, a.e. 5, l. 4.

33 Report of Petar Ignatov, chief director of Balkantourist, to the Minister of Trade regarding action measures for developing international and domestic tourism in Bulgaria, undated, TsDA, f. 310, op. 2, a.e. 262, l. 38–39.

34 The transit passengers were divided into three main categories: 69 per cent were Turkish guest workers in Western Europe (which represented more than half of all Western visitors), 19 per cent were Middle Eastern guest workers (mainly from Syria, Jordan, and Lebanon), and 22 per cent were Western holidaymakers en route to Turkey (mainly from Austria, Germany, and the UK). An additional category of incoming Westerners, accounting for around 5 per cent of the total, were on business. Notes from the Committee for State Security on the Report by the Minister of Finance and the Chairman of CDT regarding the increase of foreign-currency revenues of international tourism, 8 December 1965, TsDA, f. 136 (Council of Ministers), op. 42, a.e. 90, l. 31–32.

35 Report of Petko Todorov, chairman of CT, to the Chairman of the Committee for Economic Coordination regarding the Conception for the development of sector "International and internal tourism" during the sixth five-year plan 1971–1975, 8 August 1970, TsDA, f. 1230, op. 1, a.e. 54, l. 99.

36 Internationally, such statistical compilations were not unique. On the Yugoslav example, see Tchoukarine, "The Yugoslav Road to International Tourism," 110.

37 Protocol of National convention on questions of tourism organized by the CT, 25 April 1967, TsDA, f. 310, op. 5, a.e. 5, l. 2.

38 Decree 47 of the Council of Ministers, 7 September 1973, TsDA, f. 136, op. 56, a.e. 65, l. 2.

39 Metodiev and Stoyanov, *Balgarskite darzhavni institutsii*, 127–28.

40 Decree 56 of the Council of Ministers, 14 December 1966, TsDA, f. 136, op. 42, a.e. 61, l. 2.

41 Protocol of National convention on questions of tourism organized by the CT, 25 April 1967, TsDA, f. 310, op. 5, a.e. 5, l. 30.

42 Report of Petko Todorov, chairman of CT, to the Chairman of the Committee for Economic Coordination regarding the Conception for the development of sector "International and internal tourism" during the sixth five-year plan 1971–1975, 8 August 1970, TsDA, f. 310, op. 5, a.e. 6, l. 9.

43 Report of Petko Todorov, chairman of CT, to the Chairman of the Committee for Economic Coordination regarding the Conception for the development of sector "International and internal tourism" during the sixth five-year plan 1971–1975, 8 August 1970, TsDA, f. 1230, op. 1, a.e. 54, l. 170–71. Trade experts also recognized the activities of the CT as a form of foreign trade. See, for example, Georgi Georgiev, "Harakter i formi na monopola na vanshnata targoviya" [Character and forms of the monopoly on foreign trade], *Vanshna targoviya* 11 (1969), 13.

44 Report of Petko Todorov, chairman of CT, to the Chairman of the Committee for Economic Coordination regarding the Conception for the development of sector "International and internal tourism" during the sixth five-year plan 1971–1975, 8 August 1970, TsDA, f. 310, op. 5, a.e. 6, l. 4, 11, 43–45.

45 Ordinance 140 of the Committee for Economic Coordination, 1 April 1970, TsDA, f. 310, op. 5, a.e. 1, l. 22.
46 I am grateful to the Danish embassy in Sofia for some materials on the Danish–Bulgarian diplomatic relations.
47 Letter from the Bulgarian embassy in Copenhagen to Petar Mladenov, Minister of Foreign Affairs, 17 November 1973, TsDA, f. 1477 (Ministry of Foreign Affairs), op. 29, a.e. 1057, l. 13.
48 Letter from MFA to Balkantourist, 16 April 1959, TsDA, f. 310, op. 2, a.e. 263, l. 273.
49 Report of Petar Ignatov, chief director of Balkantourist, to the Minister of Trade regarding action measures for developing international and domestic tourism in Bulgaria, undated, TsDA, f. 310, op. 2, a.e. 262, l. 37, 40.
50 Information on the work of Section "Capitalist Countries" during the first quarter of 1960, undated, TsDA, f. 310, op. 2, a.e. 287, l. 82.
51 Price-setting of all-inclusive tours, undated, TsDA, f. 310, op. 3, a.e. 16, l. 1; Letter from Bulgarian Civil Aviation to Balkantourist, 14 May 1960, TsDA, f. 310, op. 3, a.e. 16, l. 105.
52 Table of the flight costs and the revenues from flight tickets for the all-inclusive tours in 1961, 20 February 1961, TsDA, f. 310, op. 3, a.e. 16, l. 104; Information on the development of tourist relations between Bulgaria and Denmark, 26 December 1969, TsDA, f. 1477, op. 26, a.e. 1015, l. 9.
53 Information on the political, economic, industrial and scientific-technical relations between Bulgaria and Denmark, February 1973,TsDA, f. 1244 (Council of Mutual Economic Assistance), op. 1, a.e. 6781, l. 5; Information on the implementation of Politburo Resolution 168 of from 18 April 1967 and Resolution 430 from 31 October 1967 for the development of the relations between Bulgaria and Denmark, TsDA, f. 1477, op. 26, a.e. 1015, l. 4.
54 Information on the development of tourist relations between Bulgaria and Denmark, 26 December 1969, TsDA, f. 1477, op. 26, a.e. 1015, l. 9.
55 Tourist relations between Bulgarian and Sweden, 25 May 1973, TsDA, f. 1477, op. 29, a.e. 2996, l. 5. In their analyses based on external data for global trends in tourism, foreign trade experts highlighted Denmark, Sweden and Switzerland as having the highest standards of living and highest spending on tourism. Emil Georgiev, "Sastoyanie i perspektivi na zapadnoevropeyskiya turisticheski pazar" [Status and prospects on the Westeuropean tourist market], *Vanshna targoviya* 3 (1969), 12.
56 Report of Petko Todorov, chairman of CDT, to Todor Zhivkov, Chairman of the Council of Ministers, regarding some problems of the development of international tourism in Bulgaria, 7 October 1966, TsDA, f. 136, op. 42, a.e. 61, l. 69.
57 Protocol of National convention on questions of tourism organized by the CT, 25 April 1967, TsDA, f. 310, op. 5, a.e. 5, l. 4.
58 Information on the political relations between Bulgaria and Denmark, undated, TsDA, f. 1477, op. 25, a.e. 867, l. 113–14.
59 Letter from the Council of Ministers, 6 June 1969, TsDA, f. 1477, op. 26, a.e. 1020, l. 10.
60 Agreement on suppression of visas between Denmark and Bulgaria, 15 August 1967, TsDA, f. 1477, op. 25, a.e. 880, 1–5.
61 Letter from the Minister of Foreign Affairs and the Chairman of CT to the chargé d'affaires in Denmark, 11 April 1973, TsDA, f. 1477, op. 29, a.e. 1058, l. 2.
62 Program for enlargement of commercial ties, industrial and scientific – technical cooperation between Bulgaria and Denmark, undated, TsDA, f. 1477, op. 26, a.e. 1015, l. 21.
63 Ibid., 20.
64 Information on the development of tourist relations between Bulgaria and Denmark, 26 December 1969, TsDA, f. 1477, op. 26, a.e. 1015, l. 9.
65 Letter from the Bulgarian embassy in Oslo to the Ministry of Foreign Affairs, 23 August 1968, TsDA, f. 1477, op. 25, a.e. 867, l. 221–22; Information on the

development of tourist relations between Bulgaria and Denmark, 26 December 1969, TsDA, f. 1477, op. 26, a.e. 1015, l. 10.

66 Report of Petko Todorov, chairman of CT, to the Chairman of the Committee for Economic Coordination regarding the conception for the development of sector "International and internal tourism" during the sixth five-year plan 1971–1975, 8 August 1970, TsDA, f. 310, op. 5, a.e. 6, l. 166.

67 Report on the work of the representatives of the CT abroad in 1968 and their main tasks for 1969, undated, TsDA, f. 310, op. 5, a.e. 110, l. 5–10.

68 Information on the development of tourist relations between Bulgaria and Denmark, 26 December 1969, TsDA, f. 1477, op. 26, a.e. 1015, l. 10.

69 Letter from the Bulgarian embassy in Copenhagen to Petar Mladenov, Minister of Foreign Affairs, 17 November 1973, TsDA, f. 1477, op. 29, a.e. 1057, l. 13.

70 Report on the work of the representatives of the CT abroad in 1968 and their main tasks for 1969, undated, TsDA, f. 310, op. 5, a.e. 110, l. 22–23.

71 Memo on a meeting with Folketurist signed by the CT representative and the chargé d'affaires, 7 September 1970, TsDA, f. 1477, op. 27, a.e. 1115, l. 4–6.

72 Letter from the deputy minister of foreign affairs to the general director of the Bulgarian airlines Balkan, 12 September 1969, TsDA, f. 1477, op. 25, a.e. 2666, l. 7; Letter from the Bulgarian Civil Aviation to the Ministry of Foreign Affairs, 29 September 1969, TsDA, f. 1477, op. 25, a.e. 2666, l. 9; Letter from the CT to the deputy minister of foreign affairs, 15 October 1969, TsDA, f. 1477, op. 25, a.e. 2666, l. 11. On the operations of Folkturist in Eastern Europe, see Sune Bechmann Pedersen, "Eastbound Tourism in the Cold War: The History of the Swedish Communist Travel Agency Folkturist," *Journal of Tourism History* 10, no. 2 (2018).

73 Letter from the Bulgarian embassy in Copenhagen to MFA, 11 January 1972, TsDA, f. 1477, op. 28, a.e. 1102, 1–7.

74 Information on the political, economic, industrial and scientific-technical relations between Bulgaria and Denmark, February 1973, TsDA, f. 1244, op. 1, a.e. 6781, l. 8.

75 Protocol of the Second session of the MBDC, 22 August 1969, TsDA, f. 1477, op. 26, a.e. 1020, l. 15.

76 Memo signed by the trade representative and the first secretary at the Bulgarian embassy in Copenhagen, 25 January 1971, TsDA, f. 1477, op. 27, a.e. 1107, l. 19.

77 Information on some issues of Bulgarian – Danish economic cooperation, undated, TsDA, f. 1477, op. 27, a.e. 1107, l. 25.

78 Information regarding the meeting on 29–30 June [1971] between representatives of the Bulgarian and Danish parties at the MBDC, TsDA, f. 1477, op. 27, a.e. 1107, l. 145; Letter from the Bulgarian embassy in Copenhagen to Vladimir Ganovski, chairman of the Bulgarian party at the MBDC, 17 September 1971, TsDA, f. 1477, op. 27, a.e. 1107, l. 157.

79 Report of Petko Todorov, chairman of CDT, to Todor Zhivkov, Chairman of the Council of Ministers, regarding some problems of the development of international tourism in Bulgaria, 7 October 1966, TsDA, f. 136, op. 42, a.e. 61, l. 63.

80 Cable from the Bulgarian embassy in Copenhagen to the MFA, 10 June 1969, TsDA, f. 1477, op. 25, a.e. 889, l. 4.

81 The Bulgarian tourist representative in Copenhagen sent a sample of translated publications from *Ekstra Bladet* (7 June 1969; also two separate materials from 9 June 1969) and *Berlingske Tidende* (9 June 1969). Letter from the CT representative in Copenhagen to the deputy chairman of the CT, 12 June 1969, TsDA, f. 1477, op. 25, a.e. 889, l. 6–11.

82 Letter to the Bulgarian embassy in Copenhagen, 3 July 1969, TsDA, f. 1477, op. 25, a.e. 889, l. 5.

83 Cable from the Bulgaria embassy in Stockholm to the CT and MFA, 19 June 1969, TsDA, f. 1477, op. 25, a.e. 2666, l. 4.

84 Cable to the Bulgarian embassy in Stockholm, undated, TsDA, f. 1477, op. 25, a.e. 2666, l. 5.
85 Letter from the Bulgarian embassy in Copenhagen to the MFA and CT, 10 May 1969, TsDA, f. 1477, op. 25, a.e. 897, l. 4.
86 For a comparison, that year, Bulgaria received 216,974 visitors from West Germany, its largest tourist provider; 74,911 from the UK; 62,463 from France; 43,074 from Italy; and 17,850 from Belgium. Yet, compared to these countries, the Scandinavian inflow stood out with its higher share of organized tourists and a much lower share of transit passengers. Report on the work of the CT representatives in 1975 on the fulfilment of the hard-currency plan of international tourism and the main tasks for 1976, TsDA, f. 1230, op. 1, a.e. 64, l. 22–33.
87 Letter from the Bulgarian embassy in Copenhagen to Tano Tsolov, Chairman of the Commission for Economic and Scientific-Technical Cooperation, 4 October 1975, TsDA, f. 1244, op. 1, a.e. 8100, l. 95.
88 Transcript translated from German (from *Stern* magazine, no. 23, 26 May 1976), TsDA, f. 1244, op. 1, a.e. 8142, l. 34.
89 Letter from the Bulgarian embassy in Copenhagen to Georgi Pavlov, deputy chairman of the Commission for Economic and Scientific-Technical Cooperation, 10 June 1976, TsDA, f. 1244, op. 1, a.e. 8142, l. 33.
90 Elitza Stanoeva, "Bulgaria's 1,300 Years and East Berlin's 750 Years: Comparing National and International Objectives of Socialist Anniversaries in the 1980s," *CAS Working Paper Series* 9/2017, 29–30.
91 CSCE, *Conference on Security and Co-operation in Europe Final Act* (Helsinki: 1975), 32, 41.
92 Vacationland Bulgaria: The Tourist Boom, J. V. Storojev, 24 February 1966, HU OSA 300-8-3-15923, 1.
93 Martin Ivanov, *Reformatorstvo bez reformi: Politicheska ikonomiya na balgarskiya komunizam 1963–1989* [Reformation with no reforms: The Political Economy of Bulgarian Communism 1963–1989] (Sofia: Ciela, 2008).

References

Bechmann Pedersen, Sune. "Eastbound Tourism in the Cold War: The History of the Swedish Communist Travel Agency Folkturist." *Journal of Tourism History* 10, no. 2 (2018): 130–45.
Bracewell, Wendy. "Adventures in the Marketplace: Yugoslav Travel Writing and Tourism in the 1950s – 1960s." In *Turizm: The Russian and East European Tourism under Capitalism and Socialism*, edited by Anne E. Gorsuch and Diane P. Koenker, 248–65. Ithaca: Cornell University Press, 2006.
Carter, Frank W. "Bulgaria." In *Tourism and Economic Development in Eastern Europe and the Soviet Union*, edited by Derek R. Hall, 220–35. London: Belhaven Press, 1991.
CSCE. *Conference on Security and Co-operation in Europe Final Act*. Helsinki: 1975. www.osce.org/helsinki-final-act?download=true.
Duda, Igor. "Workers into Tourists: Entitlements, Desires, and the Realities of Social Tourism under Yugoslav Socialism." In *Yugoslavia's Sunny Side: A History of Tourism in Socialism*, edited by Hannes Grandits and Karin Taylor, 33–68. Budapest: Central European University Press, 2010.
Georgiev, Borislav. "Arhitekturno-gradoustroystveniya konkurs za nov chernomorski kompleks kray s. Primorsko" [Architectural and Urban Planning Competition for a New Black Sea Resort Complex Near the Village of Primorsko]. *Arhitektura* 5 (1960): 26–29.

Georgiev, Emil. "Sastoyanie i perspektivi na zapadnoevropeyskiya turisticheski pazar" [Status and prospects on the Western European tourist market]. *Vanshna Targoviya* 3 (1969): 10–13.

Georgiev, Georgi. "Harakter i formi na monopola na vanshnata targoviya" [Character and Forms of the Monopoly on Foreign Trade]. *Vanshna Targoviya* 11 (1969): 10–13.

Guentcheva, Rossitza. "Mobile Objects: Corecom and the Selling of Western Goods in Socialist Bulgaria." *Balkan Studies* 45, no. 1 (2009): 3–28.

Hall, Derek. "From 'Bricklaying' to 'Bricolage': Transition and Tourism Development in Central and Eastern Europe." *Tourism Geographies* 10, no. 4 (2008): 410–28.

Hristov, Todor. "The Dry Cleaning of a Socialist Economy: The Bulgarian Debate on the Concept of Service Industry in the Context of the 1967 Economic Reforms." Paper presented at Reforming Socialism: Aims and Efforts Before and After 1968, European University Institute, Florence, 25–27 October 2018.

Ivanov, Martin. *Reformatorstvo bez reformi: Politicheska ikonomiya na balgarskiya komunizam 1963–1989* [Reformation with no reforms: The Political Economy of Bulgarian Communism 1963–1989]. Sofia: Ciela, 2008.

Kozovoi, Andrei. "The Way to a Man's Heart: How the Soviet Travel Agency 'Sputnik' Struggled to Feed Western Tourists." *Journal of Tourism History* 6, no. 1 (2014): 57–73.

Light, Duncan. "'A Medium of Revolutionary Propaganda': The State and Tourism Policy in the Romanian People's Republic, 1947–1965." *Journal of Tourism History* 5, no. 2 (2013): 185–200.

Metodiev, Veselin and Stoyanov, Lachezar. *Balgarskite darzhavni institutsii 1879–1986* [Bulgarian state institutions 1879–1986]. Sofia: D-r Petar Beron, 1987.

Petrescu, Dragoş. "Closely Watched Tourism: The Securitate as Warden of Transnational Encounters, 1967–9." *Journal of Contemporary History* 50, no. 2 (2015): 337–53.

Salmon, Shawn. "Marketing Socialism: Inturist in the Late 1950s and Early 1960s." In *Turizm: The Russian and East European Tourism under Capitalism and Socialism*, edited by Anne E. Gorsuch and Diane P. Koenker, 186–204. Ithaca: Cornell University Press, 2006.

Stanoeva, Elitza. "Sotsialisticheskata targoviya v Balgariya (1954–1963): Ideologiya, distsiplina i marketing" [Socialist trade in Bulgaria (1954–1963): Ideology, Discipline and Marketing]. *Sociological Problems*, nos. 3–4 (2015): 228–49.

———. "Bulgaria's 1,300 Years and East Berlin's 750 Years: Comparing National and International Objectives of Socialist Anniversaries in the 1980s." *CAS Working Paper Series* 9 (2017): 3–40.

Taylor, Karin and Grandits, Hannes. "Tourism and the Making of Socialist Yugoslavia: An Introduction." In *Yugoslavia's Sunny Side: A History of Tourism in Socialism*, edited by Hannes Grandits and Karin Taylor, 1–30. Budapest: Central European University Press, 2010.

Tchoukarine, Igor. "The Yugoslav Road to International Tourism: Opening, Decentralization, and Propaganda in the Early 1950s." In *Yugoslavia's Sunny Side: A History of Tourism in Socialism*, edited by Hannes Grandits and Karin Taylor, 107–38. Budapest: Central European University Press, 2010.

———. "Yugoslavia's Open-Door Policy and Global Tourism in the 1950s and 1960s." *East European Politics & Societies* 29, no. 1 (2015): 168–88.

Walton, John. "Preface: Some Contexts for Yugoslav Tourism History." In *Yugoslavia's Sunny Side: A History of Tourism in Socialism*, edited by Hannes Grandits and Karin Taylor, ix–xxii. Budapest: Central European University Press, 2010.

2 The lure of capitalism

Foreign tourists and the shadow economy in Romania, 1960–1989

Adelina Stefan

In 1964, a British newsreel promoting Romania as a tourist destination described the country as a blend of tradition and modernity, which despite its location beyond the Iron Curtain displayed a capitalist mentality.[1] Romanian tourist propaganda sent a similar message when advertising Romania in capitalist countries.[2] In 1976, *Vacances en Roumanie*, a Romanian tourist magazine published abroad, invited Western tourists to spend their holidays on the Romanian "Riviera" of the Black Sea. "Roulette, jazz, beauty contests, night shows, music, projections, and cocktails" were all part of the holiday package that was supposed to render Western tourists productive for the rest of the year.[3] Yet, despite this tourist promotion, when it came to encouraging market economy practices in reality, the Romanian socialist government was less enthusiastic. This chapter thus examines the tension between the socialist regime's goal of attracting Western tourists with their coveted hard currency and its fear of capitalist "contamination" at the everyday level. More than anything else, international tourism exposed socialist society to Western eyes – and capitalist consumption patterns to the Romanian public.

Here, I explore the politics of the Romanian socialist state regarding international tourism, and then look at the economic interaction between foreign tourists and Romanian citizens, to gauge the extent to which these contacts eluded the state, before examining the ways in which consumer culture in Romania was reshaped by contact between tourism workers and Western tourists. I argue that the direct contacts between Romanians, and especially tourism workers, and Western tourists triggered significant changes in their taste and dress, and stimulated their entrepreneurial mentality. Most of these changes were seen on the Black Sea coast, where Western tourists predominated, and in the Transylvanian towns of Sibiu and Brasov, home to a sizeable German community.

This chapter adds to the growing literature on the porousness of the Iron Curtain and the role of international tourism in that.[4] The cultural turn in Cold War studies shifted the discussion from "the culture of the Cold War" to "Cold War cultures," as Rana Mitter and Patrick Major have it;[5] that is, from diplomatic and political relations between two nominally divergent blocs to the meanings associated with these relations and their impact at the everyday level. The question of international tourism has only recently been raised in this conversation. Recent studies by Anne Gorsuch, Diane Koenker, Igor Tchoukarine, and Sune Bechmann

Pedersen point to the role of tourism in promoting mobility in the Cold War, either within the socialist bloc or between the socialist East and the capitalist West.[6] The present study thus supplements the current literature on tourism by focusing on the effects of international tourism on ordinary lives during the Cold War, an aspect that current works on tourism have overlooked.

The Cold War was a lived experience for citizens in both the socialist East and capitalist West, and while international tourism facilitated direct contact between them it still begs the question of the extent to which it consolidated or dismantled the official rhetoric about the 'other camp', whether the communist regime in Romania or the tourists' capitalist home countries. How far did international tourism help ordinary citizens in both Romania and the capitalist West profit from the political and economic divisions between Eastern and Western Europe, if at all?

The politics of international tourism and its limitations

In the early 1960s, the socialist state of Romania was increasingly interested in welcoming foreign tourists, especially those from the capitalist countries.[7] The construction of new modern beach resorts on the Black Sea from the late 1950s onwards helped attract Western clients. The number of foreign tourists increased from 100,000 in 1960 to about 6 million in the mid-1970s. In the early 1980s, the number of foreign tourists peaked at 7 million annually. Although only 35–40 per cent of all tourists came from Western capitalist countries (or 'developed countries' as Romanian official rhetoric had it), they brought far greater revenue to the Romanian economy than did tourists from socialist countries.[8]

Tourist collaboration among the socialist countries had begun in the 1950s. A summit in Varna in 1955 staked out the general principles, and in 1957 the national tourist authorities of the Council for Mutual Economic Assistance (Comecon) member states held their first conference in Carlsbad in Czechoslovakia to discuss the matter in greater detail.[9] At first, the discussions focused on international tourism within the socialist bloc, but from the 1960s onwards the socialist countries also sought to develop international tourism across the East–West divide.[10]

A fourth meeting of tourist organisations from the socialist countries marked a change in the Eastern bloc's tourist policy. The summit took place in 1961 in Moscow and included participants not only from the Soviet Union and Eastern Europe, but also from Mongolia, North Korea, and North Vietnam. The second point on the agenda referred to the "importance of developing international tourism between socialist and capitalist countries as a means of popularising the accomplishments of socialist regimes and of counterattacking the unfriendly imperialist propaganda towards socialist countries."[11] The next point on the agenda stated that tourist relationships between socialist and capitalist countries should start from the idea that socialist states "could be less expensive and more attractive tourist destinations."[12] The meeting also emphasised that socialist countries should find ways to promote themselves on the capitalist countries' tourist market.[13] During this meeting, Romania signed tourist agreements for 1962 with Intourist (USSR), Orbis

(Poland), Čedok (Czechoslovakia), and Ibusz and Expres (Hungary). However, in the early 1960s, Romania was not the strongest voice when it came to tourist relationships with capitalist countries. At the meeting in Moscow, for example, Romania's representatives presented a report on "recreational tourism" and the prospects for its development within the socialist bloc, because this was Romania's main priority.[14] Its border-crossing policies closely reflected this stance, since in 1964 only tourists from socialist countries could travel without a passport to Romania, while in Bulgaria visitors with pre-paid vouchers were already able to receive on-the-spot visas without having to declare the currencies and amounts they were carrying on their arrival or departure.[15]

But change was underway. In 1964, the Romanian Council of Ministers decided to send a number of tourism specialists to France to receive training in hotel and restaurant management and become acquainted with French cuisine. In 1966, a report by the Ministry of Foreign Trade and International Cooperation asked Oficiul Naţional de Turism Carpaţi (ONT Carpaţi), the Romanian Tourist Office, to attract tourists from West Germany and the Scandinavian countries, as these looked to be the most promising markets. For the same financial reasons it also asked ONT Carpaţi to pay more attention to individual tourism as opposed to package tours.[16] In 1967, to boost international tourism with Western countries, Romania had already abolished the visa regulations that required potential tourists from capitalist countries to visit the Romanian embassy at home, and visas became a simple formality, automatically granted at the border.[17] Alongside the liberalisation of travel there was institutional consolidation. Also in 1967, the ONT Carpaţi, previously under the direction of the Ministry of Foreign Trade and International Cooperation, became a stand-alone institution similar to a ministry under the supervision of the Council of Ministers.[18]

In addition to sending tourist workers to be trained in Western countries, welcoming Western tourists, and creating an institutional framework, a new definition of tourism started to crystallise at the end of the 1960s. It now encompassed an economic dimension; tourism ceased to be simply a recuperative activity that improved Romanian workers' physical condition, and henceforth became a set of services designed to meet the needs of potential consumers. Oskar Snak, a top official in the Ministry of Tourism and a scholar of tourism, explained, "From an economic and social point of view, the development of tourism refers to the population's growing demands for better access to tourist services and consumer goods, which in the end stimulates both production and consumption." Furthermore, Snak emphasised that the growing number of "foreign visitors is beneficial for the development of certain tourist areas and of the Romanian economy in general."[19]

Yet, international tourism seemed to only partially meet the economic expectations of the Romanian government. Despite the liberalisation of travel and the market-oriented development of tourism, the total income from international tourism in Romania in 1982 was just 1.4 per cent of GDP, below the world average income of 3.4 per cent of GDP.[20] In fact, in 1975 Romania earned only $132 million from international tourism, while in 1980 tourism revenues climbed to $324 million only to plummet to $176 million in 1988.[21] Corneliu Mănescu, a member of the Executive Bureau of the Central Committee of the Romanian Communist Party and president of the UN General Assembly in 1967–1968, explained the limited success of international tourism in Romania as follows:

> We cannot compare ourselves with the Dalmatian Coast, we have abolished the visas but this thing did not bring too many tourists. We have to make propaganda, to build tourist circuits, to understand that tourism does not mean only the hotel or places such as Eforie and Mangalia, but we can develop tourism in other places too.[22]

The lack of flexibility in the design of international tourism was apparently the most important challenge for socialist officials. One possible explanation for this inflexibility was the structure of the planned economy, which did not leave much room for adjustments over the year, but this alone cannot explain the relatively poor performance of international tourism in Romania.[23] In fact, the most important limitation stemmed from the international market itself. In the late 1970s, the world energy crisis and its subsequent inflation in the late 1970s, as well as the return of the Cold War rhetoric and the war in Afghanistan influenced the choices of Western tourists who could not afford to take extended vacations abroad, or did not find it safe enough. Destinations beyond the Iron Curtain like socialist Romania were affected by these geopolitical developments, and the number of Western tourists plummeted in the 1980s. At the same time the Romanian government, after having prioritised international tourism in the 1970s, began to cut investment in tourist infrastructure in the 1980s and even limited the import of foodstuff for international tourism, which was considered too expensive.[24]

Figure 2.1 Postcard showing new hotels in Mamaia, Romania, in 1961

Source: In author's possession

Consuming socialism through international tourism

The Romanian communist regime was critical of conspicuous consumption among its own citizens.[25] However, the regime displayed a different attitude when it came to foreign tourists. In their case, it wanted to encourage their consumption when they were on holiday in Romania, for instance through the opening in 1964 of COMTURIST shops, which specialised in selling merchandise to foreign tourists. In 1969, the revenue generated by tourist shops amounted to $2.6 million, but the figure did not impress the regime.[26] Compared to the income generated by other countries in the socialist bloc, Romania was lagging behind. Bulgaria reportedly earned $4.5 million and Czechoslovakia no less than $45 million.[27] The Executive Bureau of the Central Committee of the Romanian Communist Party therefore complained that the ONT Carpați had failed to achieve its goal, and it looked for ways to improve revenues from foreign tourism.[28]

The proposed solutions ranged from "making available a large array of merchandise from both internal and external production such as cosmetics, food, cars, apartments, construction materials, and medicines" to selling those goods at reasonable prices (if possible, at lower prices than in the tourists' home countries).[29] The main problem seemed to be the difficulty of adjusting to consumer demand, which the proposals for improvement did not address in any detail.[30]

Although the number of tourists increased steadily until the early 1980s, it was still below the planned number, and not enough to fill the tourist facilities built to accommodate them.[31] In 1966, communist officials were dissatisfied because the seaside occupancy rate was at only 60–70 per cent during the peak season.[32] According to ONT Carpați officials, the problem was a lack of adequate services and tourist personnel. This was why they asked for an investment of 3 billion lei for the 1966–1971 period and an increase in the number of tourism workers. The demand was met with scepticism by some members of the government, including the president of the Council of Ministers, Ioan Gheorghe Maurer, yet ultimately it was approved.[33] What is more, in 1966 the Council of Ministers and the Central Committee of the Romanian Communist Party approved a new plan for the "systematization" of the Romanian seaside, which called for the building of a new resort for foreign (Western) tourists on the southern part of the coast.

Though the Romanian socialist state wanted to have more Western tourists, it was also preoccupied with policing the interactions between Western tourists and Romanians, because it saw the former as a possible source of "anti-socialist" contamination. What worried the communist government were the informal economic practices that flourished between foreign tourists and Romanians. Romania mostly focused on industrial development, and allocated more than 50 per cent of its investment to the industrial sector at the expense of agriculture and services. Little attention was paid to domestic consumption, and the availability of consumer goods remained patchy, with the exception of a short period in the late 1960s.[34] Moreover, goods made in capitalist countries were far out of the reach of ordinary citizens. But the arrival of Western tourists and their access to the coveted tourist shops provided an alternative outlet for consumer goods for Romanian citizens. The state did not encourage these lucrative interactions as they compromised the image of the socialist regime, and revealed its inability to cater for its citizens.

Consequently, the Securitate, the infamous secret police, periodically instructed tourism workers, who were more likely to come into contact with tourists from capitalist countries, not to accept gifts or engage in economic activities with foreign tourists.[35] To 'justify' its actions, the secret police used patriotic rhetoric as it warned citizens that such economic exchanges were a cover for espionage. Securitate's "Note on the counterrevolutionary preparations of tourism workers from Sibiu County" from 1974 told tourism workers to report foreign tourists' suspicious behaviour within 24 hours, or even act themselves if they believed that those individuals could endanger national security.[36] The note warned tourism workers that on various occasions foreign tourists had taken advantage of tourist employees' weaknesses and offered them presents.

The behaviour the Securitate's note referred to stretched from relatively small infractions to complicated networks of foreign or Romanian currency smuggling. A common practice especially for smaller hotels was to check tourists in illegally or to encourage prostitution. A Securitate report gives detailed examples about such activities: "We consider damnable the deed of T.I. and V.V., receptionists at Saliște Inn, who illegally checked in numerous individuals, including foreign tourists. They erratically registered them, after which they misappropriated the payment, and falsified the hotel records."[37] In another situation, which also took place at Saliște Inn, T.I. "the director of the inn, instead of helping the militia, allowed the prostitutes to escape through the back door."[38]

The presence of foreign tourists gave tourism workers access to various material resources. For example, it helped them acquire coveted foreign currency. Foreign currency was a state monopoly, and possessing only a few dollars or Deutschmarks could land a Romanian citizen in prison.[39] Yet smuggling hard currency was a daily occurrence. One such case is the example of a watchman at Lebăda Cottage, in Tulcea County in the Danube Delta, who "gave foreign visitors rides with the cottage boat for which he charged them in Deutschmarks, or invited tourists to have dinner at his house where he cooked fish dishes, thus gaining their trust."[40] Through his daily contact with foreign tourists, this otherwise rather anonymous blue-collar worker in the Danube Delta gained access to hard currency, which he could later use to buy goods from the tourist shops.

Currency restrictions also applied to foreign citizens when it came to Romanian lei. They were allowed to bring any quantity of hard currency into the country, but they were not allowed to take Romanian lei out.[41] A 1967 tourist guidebook in English advised tourists to exchange their remaining Romanian money at "a bank, bureau de change, or the nearest National Tourist Office" before leaving Romania.[42] Yet this was not economically advantageous for foreign tourists as the official rate disproportionally favoured the state.[43] When arriving in Romania, tourists could exchange money at the "exchange bureau of the National Bank of Romania, in big hotels, at airports, ports, and railway stations, as well as at all National Tourist Office agencies and branch offices in Bucharest and throughout Romania,"[44] where tourists also had the option to exchange travellers' checks. To prove to customs officials that they changed all their lei when leaving the country, foreign tourists were instructed to save all their receipts.[45]

In spite of these measures, the smuggling of both Romanian and foreign currencies was a routine activity from the 1960s to the 1980s. A 1965 secret police report noted that, "The cases that we have discovered prove that such illegal transactions [smuggling] involve both foreign and Romanian currency."[46] Not surprisingly, Western tourists were active participants in the smuggling of Romanian currency, as it was a profitable business. The same 1965 Securitate note emphasised that:

> Some foreign citizens purchased and took out of the country Romanian currency with the purpose of selling it abroad at a better price or exchanging it for other currencies. The currency exchange took place not only at oversees exchange offices, but also between private citizens. These individuals intend to visit our country and need Romanian lei.[47]

Austrians and West Germans who had emigrated from Romania were featured large in such transactions. This was the case with Ernst F. from Austria, who coordinated illegal currency exchanges with his brother Richard F., a Romanian citizen living in Sibiu (Hermannstadt). In one case, Ernst F. carried 200,000 lei (the equivalent of $12,000) across the Romanian and Hungarian borders.[48] The case was discovered when a spiteful neighbour informed on him. When finally caught by the Romanian authorities, F. told them that he "only exchanged 80,000 lei and brought the rest of the money back to invest it in jewellery, as the currency exchange business was not that profitable."[49]

This was hardly an isolated case. Between 1963 and 1965, another Austrian visitor, Iosif H., sold various Western commodities, such as razors, marker pens, and tablecloths, to obtain important revenues in Romanian currency. He then used the Romanian money to buy goods from a Viennese store that accepted Romanian lei, or sold it to prospective tourists bound for Romania.[50] His ultimate goal was to exchange the Romanian money for dollars. The Viennese shop that accepted payments in lei officially sold Romanian folk artefacts, but in the background, it actually operated an efficient network of currency exchange. Clearly, the socialist state was not the only beneficiary of Romania's opening up to foreign tourists. In addition, some citizens of capitalist countries started to sell Romanian currency in the West. As a city perched between the socialist and capitalist blocs, Vienna became a very important location in this network. At the end of the 1960s, a representative of a foreign travel firm in Romania noted that "passing through Vienna I saw that there are large quantities of Romanian money that sell for 20–22 lei per one US dollar."[51] This story suggests there was a well-established network for the smuggling of Romanian lei – and that the Romanian socialist state was unable to halt it. Moreover, it shows how ordinary citizens in Romania and some citizens of capitalist countries capitalised on the East–West divisions of the Cold War and the Romanian state's restrictive policies on foreign currency exchange. This happened against the backdrop of international tourism which could afford to ignore the Iron Curtain, but was only partially to the Romanian state's advantage.

Fighting the Cold War on the Black Sea Riviera: informal relations between tourism workers and Western tourists

A complex set of relationships was established between foreign tourists and Romanians, be they tourism workers, domestic tourists, or relatives of foreign tourists living in Romania (most of them ethnic German). These networks enabled Romanian citizens to circumvent state authority and either run private enterprises in a state-socialist economy or simply get access to consumer goods that were not available in the shops.[52] At the same time, these informal relations developed into transnational networks that reached beyond the socialist camp.

The interactions between foreign tourists and Romanians ultimately fostered a different view on lifestyle and consumption among ordinary citizens in Romania. Tourism workers who came into daily contact with foreign tourists were among the Romanian citizens most affected by these changes in mentalities and lifestyles. Doina, one of my interviewees, worked as a maître d'hotel in Neptun, a holiday resort built on the Romanian Black Sea coast in the 1970s, after having started as a waitress. She recalled that, "When I came here, I thought I was in another country."[53] Her story is similar to that of many tourism workers who took advantage of the development of tourism in Romania in the 1970s, and enthusiastically poured into the newly built holiday resorts.[54] Despite the low wages, tourism workers were brought to the coastal resorts by the chance to informally trade with foreign tourists and to live in a more cosmopolitan milieu.

Tourism workers' reminiscences are an excellent source to document the ways in which consumption habits changed after taking a job in tourism on the Black Sea coast. While Doina (aged 55 at the time of the interview) began to work in the early 1970s, a period of a relative political and economic liberalisation, two interviewees (a man aged 44 and a woman aged 42) started work in the mid-1980s, at the peak of the consumer goods scarcity in Romania.[55]

As maître d'hotel in Neptun, a resort where tourists from capitalist countries predominated, Doina deems her job in tourism to have been an opportunity compared to how her life would have turned if she had stayed in her home town in Moldavia (a region in eastern Romania) and worked in a factory.[56] Just seeing how female foreign tourists dressed and behaved taught her about fashion and modern lifestyles. Yet, the economic restrictions made it difficult for Romanian citizens to buy these goods from ordinary shops. The "tourist shops," which sold goods in hard currency and were conveniently located in every major hotel along the coast, was one place where tourism workers could buy Western manufactured goods. Yet, as tourism workers could not legally own foreign money, they could not just go and buy what they wanted; they had to ask their foreign friends to do it for them. Doina recalls befriending a foreign tourist from West Germany, who agreed to buy a fleece jacket for her, which she describes as being, "A little bit more different than what other people wore."[57] Through this shopping by proxy, Doina managed to acquire the goods she was yearning for.

Ion T., a waiter at Hotel Doina in Neptun, remembers how at the end of their stay the tourists would collect money, buy things from the shop, and offer presents to every tourist employee from the maid to the receptionist. It was also usual for most tips to be in foreign currency, which Ion T. would also use to "buy" things from the hotel's shop. The shop played a central role in his recollections, as, like Doina, this was where he could access goods that were not available in ordinary shops. Although Ion T. had only a very elementary knowledge of German, he was still able to strike up contact with foreign tourists with whom he stayed in touch after 1990, when they could come and visit him at home.[58]

Alexandra N. started working in tourism when she was 14 years old. She began as a kitchen assistant and reached the peak of her career as a receptionist. She met foreign tourists on a daily basis when she worked at Caraiman, a three-star hotel in Neptun. Foreign tourists, particularly those in their 60s or 70s, liked the way she behaved towards them, and they were always curious to know about her age, family, etc. Foreign tourists were quite aware of the scarcity of consumer goods in the 1980s, and they used it to get acquainted with tourism workers – and consequently to obtain better service. Alexandra N. remembers getting tips in foreign currency that tourists "left on the table under the napkin." She used her first money to buy a tracksuit, and after she had saved for a couple of months she bought a Grundig dual cassette deck from the tourist shop with the help of foreign tourists. In most cases, the tourism workers' shopping experience was mediated by the foreign tourists, and required social skills. The presence of Western tourists helped them get the goods they wanted, and to get around the Romanian state, which sought to control the public and private lives of its citizens. At the same time, the state through its agents (and especially party officials and militia) to some degree tolerated this informal system because it supplied the goods that the official retail system could not provide.

Conclusions

As Caroline Humphrey argued in her study of personal property in socialist Mongolia, material possessions matter, and have both identity and ritualistic significance in people's lives.[59] However insignificant the gifts or goods that tourism workers received from foreign tourists, they were extremely meaningful in the context of the consumer goods shortage in the Romania of the 1970s and 1980s. For tourism workers, these goods opened a window onto a world that was not physically accessible to them, as they could not easily travel to the West. In most cases, the possession of trivial Western items had a symbolic meaning, proving their grip on that "world."

There were limits to what international tourism accomplished in Romania, compared to other socialist countries such as Yugoslavia or Czechoslovakia, but it definitely put Romania on the world tourist map. As more foreign tourists poured into the country, it became increasingly difficult for the socialist state and the Securitate to control the interactions between foreigners and Romanians, especially those who worked in the tourist sector. Tourist employees obtained privileged access to consumer goods and developed a cosmopolitan consumption pattern at odds with the official ideology of rational socialist consumption.[60]

Furthermore, the boundaries between the official and unofficial economies were blurred by the economic interactions between foreign tourists and tourism workers. By using their position in the socialist economy, tourism workers became a privileged group, and even displayed a capitalist mentality in their relationships with Western tourists. This ambiguous situation was best explained by the anthropologist Alexei Yurchak, who showed that through its agents (in this case tourism workers) the state itself overlooked socialist ideology.[61] In the long run, informal economic relationships between foreign tourists and Romanian citizens compromised the legitimacy of the socialist regime, as they not only showed the state's inability to fulfil the citizens' consumer needs, but they also honed a market mentality among tourism worker and their peers. Not surprisingly, then, tourism workers were among the first entrepreneurs in the post-1989 period.

Notes

1 Rumania 1964, www.britishpathe.com/video/rumania.
   I wish to thank Claudiu Oancea and Sune Bechmann Pedersen for commenting on this chapter at an early stage; the Central European University's Institute for Advanced Study, where I was a Humanities Initiative Fellow for a year; and for the financial support of the PanEur1970s project "Looking West: The European Socialist Regimes facing Pan-European Cooperation and the European Community," financed by the European Research Council and based at the European University Institute, Florence.
2 The news feature may have been sponsored by the Romanian government.
3 "Casinos et night clubs," *Vacances en Roumanie*, no. 49 (1976): 6.
4 For the Iron Curtain and revisiting the Cold War, see Simo Mikkonen and Pia Koivunen, eds., *Beyond the Divide: Entangled Histories of Cold War Europe* (London: Berghahn, 2015); Peter Romijn, Giles Scott-Smith, and Joel Segal, eds., *Divided Dreamworlds: The Cultural Cold War in Western and Eastern Europe* (Amsterdam: Amsterdam University Press, 2012); Annette Vowinckel, Marcus Payk, and Thomas Linderberger, eds., *Cold War Cultures: Perspectives on Eastern and Western European Societies* (New York: Berghahn, 2012); Sari Autio-Sarasmo and Katalin Miklóssi, eds., *Reassessing Cold War Europe* (London: Routledge, 2011); Gordon Johnston, "Revisiting the Cultural Cold War," *Social History* 35, no. 3 (2010): 290–307; Rana Mitter and Patrick Major, eds., *Across the Blocs: Exploring Comparative Cold War Cultural and Social History* (London: Frank Cass, 2004).
5 Mitter and Major, *Across the Blocs*, 20.
6 For travel in socialist societies and the Cold War, see Anne E. Gorsuch, *All This is Your World: Soviet Tourism at Home and Abroad* (Oxford: Oxford University Press, 2011); Anne E. Gorsuch and Diane P. Koenker, eds., *The Socialist Sixties: Crossing Borders in the Second World* (Bloomington: Indiana University Press, 2013); Igor Tchoukarine, "Yugoslavia's Open-Door Policy and Global Tourism in the 1950s and 1960s," *East European Politics & Societies* 29, no. 1 (2015): 168–88; Sune Bechmann Pedersen, "Eastbound Tourism in the Cold War: The History of the Swedish Communist Travel Agency Folkturist," *Journal of Tourism History* 10, no. 2 (2018). See also Adelina Stefan, "Vacationing in the Cold War: Foreign Tourists to Socialist Romania and Franco's Spain, 1960s–1970s" (PhD diss., University of Pittsburgh, 2016); Adelina Stefan, "Postcards Transfer across the Iron Curtain: Foreign Tourists and Transcultural Exchanges in Socialist Romania during the 1960s and 1980s," *International Journal for History, Culture & Modernity* 5, no. 1 (2017): 169–95; Adelina Stefan, "Between Limits, Lures, and Excitement: Holidays Abroad in Socialist Romania during the

1960s–1980s," in *Socialist and Post-Socialist Mobilities*, ed. Kathy Burrell and Kathrin Hörschelmann (Basingstoke: Palgrave Macmillan, 2014), 87–105.

7 National Archives of Romania (ANIC), Central Committee of Romanian Communist Party, Economic Unit, file no. 165/1981, fol. 21, Bucharest, Romania.

8 ANIC, Central Committee of Romanian Communist Party Collection, Economic Unit, file no. 244/1981, fol. 10.

9 Martin A. Garay, *Le tourisme dans les démocraties populaires européennes* (Paris: La documentation française, 1969), 35; Sune Bechmann Pedersen, "A Passport to Peace? Modern Tourism and Internationalist Idealism," *European Review* 28, no. 3 (2020).

10 On the different approaches to international tourism within the socialist bloc see Derek R. Hall, "Tourism Development in Contemporary Eastern and Central Europe: Challenges for the Industry and Key Issues for Researchers," *Human Geographies – Journal of Studies & Research in Human Geography* 5, no. 2 (2011): 5–12.

11 ANIC, Council of Ministers Collection, file number 29/1961, fol. 6.

12 Ibid., fol. 7.

13 Ibid., fol. 10.

14 Ibid., fol. 39. The Czechoslovak delegation was in charge of the report on tourism with capitalist countries.

15 "European Travel on the Increase," *World Travel – Tourisme Mondial*, no. 63 (1964): 3, 15.

16 ANIC, Central Committee of Romanian Communist Party Collection, Chancellery Section, file no. 150/1966, fol. 2.

17 Colecţia de legi şi decrete (Bucharest: Consiliul de Stat, 1967), HCM 800/1967, 281.

18 Stefan, "Vacationing in the Cold War," 58.

19 Oskar Snak, *Economia şi organizarea turismului* (Bucharest: Sport Tourism Publishing, 1976), 28–29.

20 Careba Crişan and Gheorghe Ionel, *Tehnica Operaţiunilor de Turism International* (Bucharest: Sport-Tourism, 1984), 14.

21 Derek R. Hall, "Evolutionary pattern of tourism development in Eastern Europe and the Soviet Union," in *Tourism and Economic Development in Eastern Europe and the Soviet Union*, ed. Derek R. Hall (London: Belhaven, 1991), 377.

22 ANIC, Central Committee Collection, Chancellery Unit, file no. 47/1967, fol. 24.

23 In fact, planning became one feature of Western European states as well. See United Nations Archive, Geneva, GX 10/1, 46 (4649).

24 ANIC, Popular Councils/ Organization Collection, CPCP-DOCALS, file no. 72/1989, fol. 2.

25 Pavel Campeanu, *Coada pentru hrană, un mod de viaţă* (Bucharest: Editura Litera, 1994), 185.

26 Ibid., 4.

27 ANIC, Central Committee Collection, Chancellery Unit, file no. 92/1969, fol. 3.

28 Ibid., fol. 5.

29 Ibid., fol. 6.

30 In 1965, an Austrian tourist complained that he wanted to buy a leather jacket from the shop, but he could not find one available, although the advertising to Romania highlighted such goods. See NAR, Council of Ministers Collection, file no. 154/1965, fol. 14.

31 *Romanian Statistical Yearbook* (Bucharest: National Institute of Statistics, 1990), 569.

32 ANIC, Central Committee Collection, Chancellery Unit, file no. 96/1966 fol. 43.

33 ANIC, Council of Ministers, file no. 227/1965, fol. 32. Maurer was sceptical that international tourism will pay off and thought that the plan that ONT Carpaţi presented was shallow.

34 Vladimir Pasti, quoted in Bogdan Murgescu, *România şi Europa: Acumularea decalajelor economice (1500–2010)* (Iaşi: Polirom, 2010), 341.

35 In 1977, the Central Committee of the Romanian Communist Party issued a decision about the types of presents employees in state-owned enterprises could accept from

foreigners, which included chocolate, but not cigarettes, alcohol, coffee, or electronics. See NAR, Collection Popular Councils, File no. 83/1977, fols. 241–42.

36 Archives of the Council for the Securitate Archives, Sibiu Documentary Fund, file no. 8663, vol. 21, fol. 262. A further three decrees were issued between 1970 and 1980 to regulate the relationship between Romanians and foreigners.

37 Ibid., fol. 264.

38 Ibid., fol. 265.

39 This was common to other socialist states as well. The more liberal Hungary also prohibited foreign currency possession. Hence the film *Secret House Search* (*Titkos házkutatás*, 1960), about the secret search of private house by the AVH (secret police) in order to find contraband such as US dollars or Western goods, while the owners were at a spa (394-0-1:1/1, OSA archive, Budapest, Hungary). According to Art. 37, Decree no. 210/1960 published in Official Bulletin no. 56 (1972), the failure to declare available foreign currencies could see the culprit imprisoned from six months to five years.

40 Archives of the Council for the Securitate Archives, Tulcea Documentary Fund, file no. 19661, vol. 4, fol. 55.

41 Peter Latham, *Romania: A Complete Guide* (London: Garnstone, 1967), 65.

42 Ibid., 65.

43 Richard, American tourist, university professor, personal interview, interviewed April 2016. He recalled being approached by a Romanian at the door of an ONT Carpați office and offered the chance to buy lei more cheaply than the official rate, and a similar transaction as he left Romania.

44 Latham, *Romania*, 65.

45 Ibid., 66.

46 Archives of the Council for the Securitate Archives, Brașov Documentary Fond, File no. 1877, vol. 11, fol. 46.

47 Ibid., 46.

48 Ibid., 47.

49 Ibid.

50 Ibid., 48.

51 Ibid., 50. This was a better price for tourists; in Romania, $1 was worth 14–18 lei.

52 "Blat" in the sociologist Alena Ledeneva's terms in *Russia's Economy of Favors: Blat, Networking and Informal Exchange* (Cambridge: Cambridge University Press, 1998).

53 Doina T., 55, personal interview, March 2013. The first hotels opened in Neptun in 1972. For this chapter, I interviewed three tourist workers (two women and one man).

54 Whereas in 1966 there were some 11,000 tourism workers, in 1975 the number of tourism workers increased to about 40,000. This was low compared to Yugoslavia, where 200,000 were employed in tourism (both permanent and temporary), but far higher than countries such as Spain and Italy whose tourism employees only numbered around 20,000 people. See *Economic Review of World Tourism* (Madrid: World Tourism Organization, 1976), 33.

55 I treat these interviews as what Portelli calls a "dialogic discourse" – a narrative that emerges from the dialogue between interviewee and interviewer. My interest is not only how accurately they reproduce the chronology of events, but how they express certain details, what they forget, and ultimately how they build the discourse of their own past. See Alessandro Portelli, *The Battle of Valle Giulia: Oral History and the Art of Dialogue* (Madison: University of Wisconsin Press, 1997).

56 Doina T. interview.

57 Alexandra N., 42, personal interview, March 2013.

58 Ion T., 44, personal interview, February 2013.

59 Caroline Humphrey, "Rituals of Death as a Context for Understanding Personal Property in Socialist Mongolia," *Journal of the Royal Anthropological Institute* 8, no. 1 (2002): 65–87.

60 Rationing continued throughout the communist period in Romania. Though officially ended in the early 1960s, bread continued to be rationed in some towns until the late 1960s. Rationing was reintroduced in 1979 for petrol, and then in the 1980s for most basic goods.
61 Alexei Yurchak, *Everything Was Forever Until It Was No More: The Last Soviet Generation* (Princeton: Princeton University Press, 2005).

References

Autio-Sarasmo, Sari, and Katalin Miklóssi, eds. *Reassessing Cold War Europe*. London: Routledge, 2011.

Bechmann Pedersen, Sune. "A Passport to Peace? Modern Tourism and Internationalist Idealism." *European Review* 28, no. 3 (2020).

———. "Eastbound Tourism in the Cold War: The History of the Swedish Communist Travel Agency Folkturist." *Journal of Tourism History* 10, no. 2 (2018): 130–45.

Campeanu, Pavel. *Coada pentru hrană, un mod de viață* (Bucharest: Editura Litera, 1994).

Crişan, Careba, and Gheorghe Ionel. *Tehnica Operaţiunilor de Turism International* (Bucharest: Sport-Tourism, 1984).

Economic Review of World Tourism. Madrid: World Tourism Organisation, 1976.

Garay, Martin A. *Le tourisme dans les démocraties populaires européennes*. Paris: La documentation française, 1969.

Gorsuch, Anne E. *All This is Your World: Soviet Tourism at Home and Abroad*. Oxford: Oxford University Press, 2011.

Gorsuch, Anne E., and Diane P. Koenker, eds. *The Socialist Sixties: Crossing Borders in the Second World*. Bloomington: Indiana University Press, 2013.

Johnston, Gordon. "Revisiting the Cultural Cold War." *Social History* 35, no. 3 (2010): 290–307.

Hall, Derek R. "Tourism Development in Contemporary Eastern and Central Europe: Challenges for the Industry and Key Issues for Researchers." *Human Geographies: Journal of Studies & Research in Human Geography* 5, no. 2 (2011): 5–12.

Humphrey, Caroline. "Rituals of Death as a Context for Understanding Personal Property in Socialist Mongolia." *Journal of the Royal Anthropological Institute* 8, no. 1 (2002): 65–87.

Latham, Peter. *Romania: A Complete Guide*. London: Garnstone, 1967.

Ledeneva, Alena. *Russia's Economy of Favors: Blat, Networking and Informal Exchange*. Cambridge: Cambridge University Press, 1998.

Mikkonen, Simo, and Pia Koivunen, eds. *Beyond the Divide: Entangled Histories of Cold War Europe*. London: Berghahn, 2015.

Mitter, Rana, and Patrick Major, eds. *Across the Blocs: Exploring Comparative Cold War Cultural and Social History*. London: Frank Cass, 2004.

Murgescu, Bogdan. *România şi Europa: Acumularea decalajelor economice (1500–2010)*. Iaşi: Polirom, 2010.

Portelli, Alessandro. *The Battle of Valle Giulia: Oral History and the Art of Dialogue*. Madison: University of Wisconsin Press, 1997.

Romijn, Peter, Giles Scott-Smith, and Joel Segal, eds. *Divided Dreamworlds: The Cultural Cold War in Western and Eastern Europe*. Amsterdam: Amsterdam University Press, 2012.

Stefan, Adelina. "Between Limits, Lures, and Excitement: Holidays Abroad in Socialist Romania during the 1960s – 1980s." In *Socialist and Post-Socialist Mobilities*, edited by Kathy Burrell and Kathrin Hörschelmann, 87–105. Basingstoke: Palgrave Macmillan, 2014.

————. "Vacationing in the Cold War: Foreign Tourists to Socialist Romania and Franco's Spain, 1960s – 1970s." PhD diss., University of Pittsburgh, 2016.

————. "Postcards Transfer across the Iron Curtain: Foreign Tourists and Transcultural Exchanges in Socialist Romania during the 1960s and 1980s." *International Journal for History, Culture & Modernity* 5, no. 1 (2017).

Tchoukarine, Igor. "Yugoslavia's Open-Door Policy and Global Tourism in the 1950s and 1960s." *East European Politics & Societies* 29, no. 1 (2015): 168–88.

Verdery, Katherine. *What Was Socialism, and What Comes Next*. Princeton: Princeton University Press, 1996.

Vowinckel, Annette, Marcus Payk, and Thomas Linderberger, eds. *Cold War Cultures: Perspectives on Eastern and Western European Societies*. New York: Berghahn, 2012.

Yurchak, Alexei. *Everything Was Forever Until It Was No More: The Last Soviet Generation*. Princeton: Princeton University Press, 2005.

3 Experiencing communism, bolstering capitalism

Guided bus tours of 1970s East Berlin

Michelle Standley

When it comes to the history of the Cold War and popular culture, tourism has been the elephant in the room, so present, so obvious, that it has easily been ignored.[1] Yet, the era of the Cold War marked a deepening of the connection between mass tourism and politics, in capitalist and socialist countries alike.[2] While a few pioneering works examine the links between tourism and post-war political reconstruction or between tourism and foreign policy, none have paid particular attention to tourism as part of a broader state strategy, East and West, to secure political legitimacy. Nor have they, by extension, examined the issue of how tourist-citizens engaged with such strategies, much less the question of whether or not such strategies were effective.[3]

In divided Berlin, in particular, mass tourism spanned the ideological and geographical cleft between the two states and societies. State and party authorities in East Berlin, the capital of the German Democratic Republic (GDR), as well as state and market leaders in West Berlin, a ward of the Federal Republic of Germany (FRG), both turned to tourism campaigns and to the cultivation of tourism. The authorities on both sides of the Wall viewed tourism as an extension of propaganda, and as a potential source of political legitimacy among their own citizens and among an international audience.

To help unravel the links between the growth of tourism and political legitimacy in the take-off phase of global tourism in the late 1960s and into the 1970s, I turn to East Berlin. I focus on official Reisebüro-DDR (Travel Agency GDR) bus tours in which only visitors from capitalist countries participated. Here I am interested less in the official discourse reproduced by the GDR tour guides, which I have considered elsewhere.[4] Instead I consider the more nebulous question of reception. How, I ask, did day trippers from capitalist countries encounter the East German capital? What did they see, feel, or recall while riding on a bus through the city? And how did they interpret what they saw and felt? In what ways did their visit impact their ideological commitment to capitalism and relationship to their home state's political regime? To answer these questions, I examine monthly reports produced by Reisebüro-DDR tour guides, who recorded comments that visitors from capitalist countries made during bus tours of East Berlin.

The focus of this study is capitalist visitors to socialist East Berlin, the majority of whom came from either West Berlin or the FRG. However, the limitations of the sources make it difficult to draw specific conclusions that link the recorded comments to the visitors' status or national origins. Generally, the guides did not record the guests' age, class status, gender, occupation, or even their country of origin. On occasion the reports note that a guest was a member of a bowling league or school group from the FRG or an insurance agent from the US travelling with friends, but for the most part the reports record a guest's comment without any additional information. Some inferences are possible based on the nature of the comments, such as a visitor stating, "*We* paid for the restoration of the Berlin Cathedral," making it evident that they came from the FRG. The Reisebüro-DDR statistics preserved in the Landesarchiv Berlin likewise lump all travellers from "non-socialist countries" into one category. Given these source limitations, I have chosen to maintain the guides' broad category of "visitors from capitalist countries," while keeping in mind that most came from West Berlin and the FRG.

My argument here is twofold. First, I claim that in order to understand and better evaluate the meaning that tourists attach to their perception of a tourist destination, scholars need to integrate into their analysis the experiential aspects of the tourist encounter. The point here is not to discount the power of the visual and symbolic, articulated most forcefully by scholars of cities such as Walter Benjamin, Guy DeBord, David Harvey, and Henri Lefebvre, and later made canonical in Tourism Studies by John Urry with his notion of the "tourist gaze." Rather it is to add further nuance to their claims, building on recent scholarship that stresses the emotional and experiential as important interpretive paradigms for evaluating the meaning that tourists attach to what they perceive while engaged in tourist activities. Second, based on an analysis of the material that takes the experiential into account, I would argue that leisure travel by visitors from capitalist countries to East Berlin helped secure capitalism and liberal democracy's political legitimacy at a time when both appeared to be in crisis.[5]

The rise of tourism globally and in the GDR

Amidst all the turmoil of the late 1960s, perhaps few people outside of those who worked in the tourism industry took note of the fact that the UN had declared 1967 the International Tourist Year. Tourism did not exactly need an official endorsement from the UN, however; between 1960 and 1967 international leisure travel had increased dramatically, at a rate of roughly 9.4 per cent yearly.[6]

As part of the international boom in tourism, rates of travel to East Berlin also grew during these years. There were nonetheless some additional factors that led to increased travel to East Berlin. First, after 1967, when the GDR introduced the five-day working week, more East Germans could undertake weekend excursions to their capital. Second, from the early 1970s on, more foreign visitors descended on the city. The largest contingent of foreign visitors hailed from Poland, Hungary, and Czechoslovakia, who after 1972 could enter the GDR without visas; the second largest group came from West Berlin and from the FRG, who after 1973 could enter the country with visas.[7]

Beginning in the late 1960s, party-state leaders in the GDR saw tourism in East Berlin as an important source of propaganda, as a means of bolstering their legitimacy, of unifying their population, and of convincing foreign visitors of the superiority of their system, the GDR's version of socialist modernity. With this in mind, they poured resources into building up the city's tourism infrastructure: publishing guidebooks in several languages; installing tourist information centres; producing souvenirs; and establishing restaurants and bars that would cater to visitor expectations of the authentically local. They also sought to find ways to shape tourist impressions of East Berlin by offering a variety of guided tours, installing kiosks with maps and wall texts at key sites, and disseminating published material.[8]

After the Basic Treaty came into effect in June 1973, by which the two German states agreed to recognise each other as sovereign states, FRG leaders actively encouraged their citizens to travel to the GDR, including facilitating school tours and other officially organised visits to the GDR capital. Leaders in West Berlin and the FRG promoted travel to the East as part of the policies linked to Ostpolitik. Introduced by former West Berlin mayor and later chancellor, Willy Brandt, Ostpolitik represented a shift in foreign policy, advocating for acceptance of the status quo of division and for direct engagement with the GDR, and by extension with the socialist bloc in general. West Germany's embrace of Ostpolitik mirrored the larger shift in dialogue between the Soviet Union and the US, as part of the policies of detente. There was another reason why leaders in the western half of the city and country supported travel to East Berlin. They were confident that West Berliners and West Germans who witnessed the contrast between the GDR and the FRG firsthand would return steadfast in their commitment to the capitalist state.[9]

Such official encouragement aside, both curiosity and a longing for excitement, even perceived danger, led many citizens from West Germany as well as those from other capitalist countries to visit East Berlin. Many leisure travellers wanted to see with their own eyes things such as the Wall that they had read about in the press or in spy novels such as John le Carré's *The Spy Who Came in From the Cold*, or seen depicted in films such as Guy Hamilton's *Funeral in Berlin*. This desire to experience the thrill of the Cold War as a tourist often put such visitors at odds with official tourist authorities in both East and West Berlin.[10] It also shaped perceptions of the divided city, likely leading to heightened nerves and an impulse towards an ideological, emotional reaction if not to the city – as viewed through the tour bus window – then to the official narrative of it as offered by the tour guide.

The bus tours

The number of visitors who took bus tours increased dramatically over the course of the 1970s. In the high season in 1970, May to September, an average of some 14,000 visitors toured the city on a Reisebüro-DDR-led bus tour each month.[11] By 1978 Travel Agency GDR reported that over 87,000 visitors had taken a Travel Agency GDR-led tour in August alone.[12] The bus tours represented a unique arrangement between the tourism authorities in East and West Berlin. The East German authorities granted three West Berlin-based tour companies, Berliner Bären (BBS), Severin + Kühn, and Berolina admittance to East Berlin, as long as they agreed to a prearranged tour led by an official Reisebüro-DDR tour guide.[13] Bus tour customers thus got on the bus in the western half of the city and crossed the border together into the eastern half. At the border, East German police inspected the bus and its passengers' paperwork. If everything appeared to be in order, they allowed the bus to proceed. After crossing the border, the bus stopped briefly to allow a Reisebüro-DDR guide to board and begin her narration of the city and of its political system.

The Reisebüro-DDR divided up the bus tours according to the visitors' place of origin.[14] They grouped visitors from West Berlin and West Germany together, assigning them a tour guide who was a member of the Communist Party. Their tours began at Invalidenstrasse. Visitors from other capitalist countries began their tours at Friedrichstrasse, and visitors from the GDR and other socialist countries started at Berolinastrasse.[15] By dividing up the guests in this manner the Reisebüro-DDR could avoid socialist tourists mingling with capitalists – the GDR had the persistent and justified fear that its citizens might make contact to facilitate escape or smuggling – and ensure that the guide could tailor her comments to the specific audience.

Berliner Bären (BBS), Severin + Kühn, and Berolina offered the option of a tour of the western half of Berlin and then, for an additional fee, adding on a brief tour of the eastern half. Alternatively, tourists could pay to visit only the eastern half.[16] For their part, the Reisebüro-DDR offered various tours of the GDR capital, ranging from a 75-minute tour to a 6-hour one, which they introduced in 1978. Designated "Tour 9," it included lunch, a tour of Arnimplatz (a square that was then undergoing reconstruction) and a visit to the Information Centre at the TV Tower completed in 1972, three years after the inauguration of the TV Tower itself. The Reisebüro-DDR planned to introduce more of these special tours as they proved popular and brought in a great deal of foreign currency.[17] Both the shorter and longer tours started with a stop at the Information Centre at the TV Tower, where guests watched a short film. From there they returned to the bus for a tour of parts of the city, with stops at the Soviet Memorial at Treptow Park and at HOG Zenner and Intershop, before ending at the Pergamon Museum.[18]

The tour bus constituted a unique sphere of interaction and group identity formation. It shuttled the driver, passengers, and tour guide along the streets of East Berlin in an enclosed space, a kind of sociocultural bubble on wheels. This had several consequences that shaped the interaction between the members of this temporary group and ultimately the visitors' perceptions of East Berlin as representative of the GDR and their own (re)identification with the capitalist West. First, the enclosed space necessarily made interaction between the passengers and local East Germans rare. Second, it focused the passengers' attention even more than normal on the city guide – the one person everyone on the bus had in common – adding weight to her role as a representative of the GDR. Third, it kept the

Figure 3.1 Berlin Alexanderplatz in 1964, before the square's extensive reconstruction as the showcase of "socialist modernity"

Source: © SLUB Dresden, Deutsche Fotothek, Richard Peter

passengers spatially confined to their own culture, and thus less likely to bump up against situations or persons who might have challenged their preconceived ideas about East Berlin and the socialist system.[19] The locals and the sites outside the tour bus window thus generally remained in the background, functioning as the scenery or backdrop to the action on the bus. When the bus did stop, allowing passengers to alight and briefly engage with the world outside, the group dynamics were already firmly in place.[20]

In all likelihood, bus tour passengers who might have had very little in common had they encountered one another somewhere in West Berlin or West Germany, suddenly had the common tie of being outsiders: outsiders who shared similar ideas about communism and the GDR, and who had consumed similar

West German propaganda and anti-communist popular culture. Although many came as part of a group – school groups, bowling clubs, or friends – they likely adhered more closely not only to one another but to their fellow bus tour passengers, brought together by the fact of their being Western outsiders in a foreign landscape.

It has been argued that when attempts to maintain respect or connect with a tour group fail, or if a guide is clearly perceived negatively, this may push tourists closer together, against the guide.[21] In the case of the East Berlin bus tours this was a regular occurrence. Troublesome disruptions almost always originated from an individual or individuals who were travelling as a group. City guides complained that groups travelling together were much harder to deal with and more likely to be confrontational. It was particularly difficult, city guides noted, to convince groups of the veracity of the guide's narrative – in part because many of the passengers had taken part in a "Wall Tour" of West Berlin beforehand, thus predisposing them to a negative reaction to their tour of East Berlin. Individuals travelling as part of a group, they commented, tended to be less open to new information or a perspective that ran counter to the opinions of their travelling companions.[22]

What visitors saw: dirt and grey

Many bus tour participants commented on the appearance of dirt. Some made negative comments about the appearance of the city in general, calling the streets dirty or comparing it negatively to the western half of the city. "In West Berlin," one guest stated, "everything is a lot cleaner."[23] Some complained about the Pergamon Museum, calling attention to the dirt and weeds at the front entrance and declaring its toilets filthy. It was surprising, noted one tourist, to find that the entrance to a museum that housed such treasures was so neglected. Someone, the visitor suggested, should at least use some weed killer on the front steps.[24] "This dirty square," announced another visitor, "changes my entire impression of your city."[25]

Other bus tour passengers complained about the dirty conditions at the official restaurant stop, HOG Zenner. "Zenner is really not a representative showpiece!" declared some passengers from Itzehoe, West Germany.[26] Built in 1821–22 as "The New Garden House on the Spree," Zenner is a two-story structure with a dark wood interior that shares the shore with the Spree River cruise line, the "White Fleet."[27] Many commented on Zenner's dilapidated exterior, including its damaged façade and the dirty appearance of the square in front. They found the toilets and bathroom towels grimy, and complained that the doors to the stalls would not shut. "I do not want to eat here!" exclaimed one visitor. Another announced, "It looks like crap here!"[28] On one occasion, the city guide overheard what she described as an unpleasant discussion among guests about the fact that they had caught a glimpse through an open window of the "appallingly filthy" kitchen.[29] In her monthly report from July 1978, one city guide observed that Zenner did indeed look very dirty at precisely the place where guests alighted from the tour buses. "This absolutely needs to be changed!" she concluded.[30]

The city guide's evident concern about Zenner's dirty exterior suggests that she and the passengers understood the symbolic importance of the dirt in evidence during Reisebüro-DDR tours. HOG Zenner was a state-run restaurant, after all, and an official, prearranged stop. Dirt in this context was neither rustic, of the type tourists might have encountered during a ramble in the French countryside, nor historic, of the sort they might have seen during a tour of ancient Roman ruins; this was official, "socialist dirt," a dirt whose very presence undermined the authority of the tour guide and the legitimacy of the state and system that the guide was trying to promote and defend.

In addition to voicing their impression that the streets and official stops were dirty, bus tour passengers also commented on the city's overall appearance, saying it was dilapidated in parts, grey, and sad. Many of the bus tour participants turned their attention to the war-damaged, and in some cases crumbling, buildings, or to the potholes along the route. "You still see ruins here," noted one. "How is it possible," she asked, "that you still have not rebuilt everything after thirty years?"[31] Note that the word "still" (the German *noch*) surfaces in these comments, underscoring some visitors' mental yardstick of progress against which they measured their host country. That some passengers mentioned this at all is all the more surprising considering that the Reisebüro-DDR deliberately planned a route to avoid such areas, and the city guides sought to direct their attention to the newly constructed skyscrapers lining Alexanderplatz and the renovated exteriors of the historic buildings along Unter den Linden.

Other visitors saw East Berlin as devoid of colour. "Here everything is grey on grey," observed one visitor, who then added, "It's horrible to live here. Don't you have any paint or why is [it like] that?" In this case, the visitor linked the lack of colour with a poor quality of life, equating grey with a "horrible" life. Other visitors did too. One visitor posed the rhetorical question, "Why is there so much grey and sadness in your cityscape?"[32] That many bus tour participants connected the GDR's supposed colourlessness with weaknesses in the political system is also apparent in comments such as, "It looks sad here. Long live socialism!"[33]

What they felt and how they interpreted their experience

The impression that visitors took away from their brief tour of East Berlin was based not only on what they saw but also on how they felt. For example, even if the tour bus did not drive along the Wall and the tour guide remained pointedly silent on the subject, just being in East Berlin conjured up a sense of being imprisoned. Not the site, but the perceived presence of the Wall inspired a feeling of oppression in guests. Members of a bowling league from Paderborn, West Germany, observed, "When you cross through the Wall, you have to be afraid that you will be arrested."[34] At the Treptow Memorial, a teacher from the FRG informed the city guide that whenever he visited East Berlin he had the oppressive feeling that he was in prison. Another West German visitor told the city guide that they found the border crossing depressing, "One has the feeling of being imprisoned." They then added, "We are sorry to see how you live in such unfreedom."[35] Other visitors also used on the metaphor of jail to describe their impression of the GDR. One group, primarily from Munich, told their city guide that the citizens of the GDR were behind bars: "You live in prison here and are not allowed out."[36]

Regardless of what they saw from the tour bus windows or at one of the designated stops – be it new skyscrapers or war-damaged blocks of flats – many filtered it through a pre-existing schema of progress versus backwardness. Some, for example, stated quite openly that they thought that the GDR was a "less developed" country. "The GDR," one bus tour participant remarked, "is still in the beginning phase of its economic development."[37] Such a comment is based on an implicit comparison. The GDR could only seem "less developed" in contrast to somewhere else. In this case, it is not difficult to surmise which country was the object of comparison: most bus tour passengers had started their tour in West Berlin, and would therefore have had fresh in their minds the contrast between the eastern and western sides of Berlin. Many guests explicitly compared West and East Berlin, contrasting the abundance on display on the Kurfürstendamm, with the bread lines in the East. One passenger declared, "The living standard in the FRG is higher than in the GDR. Live in the FRG for half a year and then you will stop glorifying socialism!"[38]

Moreover, guests did not see GDR successes as stemming from socialist development, but as evidence that "The GDR is finally catching up with the FRG and West Berlin."[39] Even when passengers saw something they thought might be evidence of progress, such as the restored Berlin Cathedral or the rebuilt city centre, it did not challenge their fundamental notion of the backwardness, and ultimately the inferiority, of the GDR. The rebuilding of Berlin Cathedral and the city centre, visitors noted, was only possible because of foreign assistance – either Western loans or direct assistance from the Soviets. A Berolina bus tour passenger announced to the guide, "The Berlin Cathedral can only be rebuilt because of funds from the West!"[40] On the same tour another passenger said, "You have told us a lot about your city. But do you also know how much debt the GDR has in the FRG?"[41]

Simply being in East Berlin, on a tour narrated by an official representative of the GDR, also conjured up images of East Berlin taken from the media, thus reinforcing passengers' perceptions of the GDR as backward. They frequently commented to the city guides on things they could not see – GDR's infamous queues for shops, and the poor quality and selection of consumer goods – as evidence of the GDR's supposedly non-existent development. "We cannot," one declared, "understand that people here still stand in lines."[42] Here the inclusion of "still" again underscores that bus tour participants saw queuing as a sign of backwardness, or a delayed arrival at the abundance already on offer in the capitalist West. "Why do you have lines in front of the bakeries?" some bus tour participants asked, then adding that the sight of the meagre displays in the GDR shops made them feel sorry for the citizens of the GDR.[43] Others even asked if food was still rationed, not realising that the GDR had ceased to ration food over a decade before.[44]

What is instructive about the frequent references to GDR citizens lining up for goods is that in all likelihood the guests did not actually witness any such thing during their visit to the GDR capital. Nowhere in their reports did the city guides mention anything about awkward encounters with queues, whereas they did mention their frustration at encounters with beggars and unsightly dirt. Furthermore, the Reisebüro-DDR carefully orchestrated the bus tours. The tour buses had to stick to a prescribed route, focused on the attractive parts of the city, and the Reisebüro-DDR had prearranged all of the stops. Thus, what the frequent comments about the presence of queues suggests is that there were not visible queues, but what might instead be termed "queues in the head," which bus tour passengers had seen in photographs or read about in newspapers beforehand. Their visit to East Berlin then conjured up latent images, overlaid on the actual sites, stirring the passengers' imagination. Whether or not it was true that the GDR, like other socialist countries, had queues is not relevant in this context; what matters here is that the visit to the host country, the socialist GDR, was an occasion to marry image with place, making what was once abstract very real.

In another context, tourists from capitalist countries might have viewed traces of dirt or even grey and the presence of older structures as romantic emblems from the pre-war past. At a time when the popular press was rife with complaints about the sterility of modern architecture, the eastern half of the city had the bulk of the Prussian architectural treasures. Furthermore, the 1970s also marked the advent of a growing interest in historical preservation and, by extension, heritage tourism. Nonetheless, if passengers did have such thoughts they either kept them to themselves, or the GDR tour guides did not feel the need to record them, as no such references appear in the archival record. Most likely, as the comments seem to indicate, the assumed economic and cultural superiority of the capitalist tourists, coupled with their overwhelmingly anti-communist orientation, stifled any potential impulse to view East Berlin with nostalgia.

Without further research one can only speculate what their reactions might have been a decade later, in the 1980s. By then, party-state authorities in the GDR had begun to respond to the groundswell of popular interest in the pre-war past and, by extension, in heritage tourism. Beginning in the early 1980s, they had officially rehabilitated those parts of the Prussian past that could be integrated into the general narrative of the GDR as the legitimate heir of the best of the German past. This return to a pre-war past included the previously unthinkable celebration of Prussian heroes such as Otto von Bismarck and religious ones such as Martin Luther, culminating in the complete restoration of a working-class street, Husemannstraße in Prenzlauer Berg, and an entire district, the Nikolaiviertel in the city centre, in time for the 750th anniversary celebrations of the founding of Berlin in 1987.[45]

Implicit criticism of capitalist society: a potential opening?

Comments from bus tour passengers hinted at what points of comparison were on their minds: student activism, the energy crisis, unemployment, and terrorism. In a report from 1971, some guests asked the guides why East Berlin did not have any problems with students, asking how authorities succeeded in keeping young people off the streets. In a report of July 1979, at the peak of the Second Oil Crisis, a city guide noted that there were numerous questions about the GDR's energy policy, such as "Are there conservation measures in view of a future energy policy?" and "How dependent is the GDR on oil imports?"[46] Or "How much of the oil imports come from the Soviet Union and how much from OPEC countries?" Other visitors asked about inflation ("Are there signs of inflation in the GDR?"). Guests also discussed terrorism, asking how the GDR dealt with it. One guest even expressed admiration for the state's "strong hand" in preventing terrorism and drug smuggling.[47]

According to one report, bus tour passengers who had been affected by the economic crisis in the FRG and West Berlin were more likely to express positive opinions about the GDR's social advances. This broad category included small-business owners and small farmers, students, workers, and interns who either had lost their positions or were under threat of losing them. Such visitors, the guides claimed, were also the most likely to voice criticism about social conditions in the FRG and West Berlin. One visitor from West Berlin described West Berlin as a dying city, which was gradually being abandoned by big industry. In the context of such comments, noted the report, bus tour participants even asked whether it would be possible for citizens of the FRG or West Berlin to work, study, or train in the GDR.[48]

In their reports, some city guides noted that guests who had not been to East Berlin for many years were pleasantly surprised by the GDR capital, and wanted to know how the state was able to provide social services and low prices. Questions such as these, the guides noted, gave them the opportunity to explain the oppositional principles underlying the two social systems in the FRG and GDR.[49] A report from early 1977 noted that some bus tour passengers from the FRG and West Berlin were "surprised and impressed" by the GDR's achievements. In general, according to another report, guests now showed greater curiosity about East Berlin and "various aspects of its social life."[50]

Fixated on the surface

Even the bus tour passengers with whom city guides could discuss the benefits of the GDR's version of socialist modernity showed time and again that they were only really interested in comparing the superficialities of life in the GDR with their country of origin. Guests from the FRG, the tour guides complained, wanted to limit the discussion to the contrasts in appearance "here and there."[51] Many visitors, said the report, were simply unable to contextualise these superficial differences, refusing to see the social causes that inspired such differences. Nor were they open to discussing the fundamental social processes and class contrasts in the FRG and GDR. Visitors from the FRG or West Berlin did not see the signs of economic crisis in the FRG as fundamental to the capitalist order, for example; they saw them as temporary "barren spells" or "hard times."

As the tour guides saw it, visitors from the Federal Republic and West Berlin "shared an anti-communist orientation" that the guides struggled to get past. Even when visitors from West Berlin and the FRG showed interest in East Berlin and the GDR, even when they expressed admiration for its successes and criticism of current conditions in West Berlin and the FRG, they nonetheless persisted in believing that capitalism and the FRG were ultimately preferable to what social-ism and the German Democratic Republic had to offer. While the guides encour-aged bus tour passengers to contrast the two states and systems, the guides and the passengers disagreed on the ground rules for the comparison, with the passengers remaining largely focused on contrasting appearances – the grey of the East with the colour of the West.

Conclusion

Visitors' reactions to the capital of the GDR ranged from curiosity and admiration to outright hostility and anger. Despite this, it is clear from their comments and behaviour on the bus tours that the experience of being in East Berlin inspired feelings (fear, imprisonment, superiority) and stirred memories (of past events and of images seen in the media) that had a powerful ideological impact that mere armchair travellers did not experience. Some bus tour passengers made comments that suggested that their visit to East Berlin was an occasion to question capitalism or the politics of their home country (asking about homelessness, rates of unem-ployment, and domestic terrorism). These openings were fleeting, however; most bus tour passengers seemed to have ended their tours of East Berlin confirmed in their conviction that, despite its flaws, capitalism was superior to communism.

In the 1970s, increased rates of unemployment, the oil crisis, terrorism, the aftermath of decolonisation, the student movement, and a decline in the moral authority of the US following the Vietnam War and Watergate all contributed to a sense that capitalism, liberal democracy, and the stability of the US-led global order were under threat. Tourism, however, helped salve some of those wounds and even suture potential ruptures. Just as globalisation and neo-liberalism – forces that would truly challenge the post-war order – began to emerge, "experiencing" communism on a bus tour through East Berlin served to blunt potential criticism and accept the flaws of capitalism. Witnessing the greyness, the fear, and the sad-ness reinforced a sense among capitalist day trippers that they not only lived on the right side of the Wall, but on the right side of history. It was with a sigh of relief that many of the bus tour passengers crossed back over the border into the West, happy to see the neon, the gaudy wares, the raucous street life, and general urban clamour on display in West Berlin, the showcase of capitalism next door.

Notes

1 Seminal works on divided Germany and popular culture include Uta Poiger, *Jazz, Rock, Rebels: Cold War Politics and American Culture in Divided Germany* (Berke-ley & Los Angeles: University of California Press, 2000); David F. Crew, *Consum-ing Germany in the Cold War* (Oxford: Berg, 2003); and Paul Betts and Katherine Pence, eds., *Socialist Modernity: East German Everyday and Cultural Politics* (Ann Arbor: University of Michigan Press, 2008). The few works on the history of post-war tourism in the FRG, the GDR, and Berlin include Scott Moranda, *The People's Own Landscape: Nature, Tourism, and Dictatorship in East Germany* (Ann Arbor:

University of Michigan Press, 2014); Rudy Koshar, *German Travel Cultures* (New York: Berg, 2000); Hasso Spode, ed., *Zur Sonne, zur Freiheit! Beiträge zur Tourismus-geschichte* (Berlin: W. Moser Verlag für Universitäre Kommunikation, 1991); Hasso Spode, ed., *Goldstrand und Teutonengrill: Kultur- und Sozialgeschichte des Tourismus in Deutschland 1945–1989* (Berlin: W. Moser Verlag für Universitäre Kommunikation, 1996); Cord Pagenstecher, *Der Bundesdeutsche Tourismus: Ansätze zu einer Visual History: Urlaubsprospekte, Reiseführer, Fotoalben 1950–1990* (Hamburg: Kovac, 2003); Claire Colomb, *Staging the New Berlin: Place Marketing and the Politics of Urban Reinvention Post-1989* (London: Routledge, 2011).

2 After the collapse of the socialist bloc and the exponential interest in cultural history, many scholars of the Cold War turned to the field of popular culture. While differing in their interpretations, what all these works shared was the premise that the Cold War marked a significant tightening at the intersection of high politics and daily life. Capitalist and socialist states, such works claimed, turned to cultural consumption as an extension of their military strategies and propaganda, pushing states and their citizens into ever-closer interdependence and heightening the expectations of both parties. Key early works include Stephen J. Whitfield, *The Culture of the Cold War* (Baltimore: Johns Hopkins University Press, 1991); Walter L. Hixson, *Parting the Curtain: Propaganda, Culture, and the Cold War: 1945–1961* (New York: St Martin's, 1997); Rana Mitter and Patrick Major, eds., *Across the Blocs: Cold War Social and Cultural History* (London: Frank Cass, 2004).

3 While no one work addresses the link between the Cold War and the rise of tourism globally, historians have addressed it in the capitalist and socialist spheres of influence respectively, for example Christopher Endy, *Cold War Holidays: American Tourism in France* (Chapel Hill: University of North Carolina Press, 2004) and Anne E. Gorsuch and Diane P. Koenker, eds., *Turizm: The Russian and East European Tourist Under Capitalism and Socialism* (Ithaca: Cornell University Press, 2006).

4 For the official narrative and the training and supervision of the tour guides, see Michelle Standley, "The Cold War Traveler: Mass Tourism in Divided Berlin, 1945–1979" (PhD diss., New York University, 2011), chs. 4 & 5, *passim*.

5 For the socialist states turning to tourism to help bolster their legitimacy, see Michelle Standley, "'Here Beats the Heart of the Young Socialist State': 1970s East Berlin as Socialist Bloc Tourist Destination," *Journal of Architecture* 18, no. 5 (2013), 683–98 where I conclude that from the early seventies at the latest the socialist bloc states and their citizens implicitly viewed tourism as an entitlement and an extension of the "socialist social contract."

6 Marie-Françoise Lanfant, "International Tourism, Internationalization and the Challenge to Identity," in *International Tourism: Identity and Change*, ed. Marie-Françoise Lanfant, John B. Allcock, and Edward M. Bruner (London: SAGE, 1995), 27.

7 M. E. Sarotte, *Dealing with the Devil: East Germany, Détente and Ostpolitik, 1969–1973* (Chapel Hill: University of North Carolina Press, 2001),72, 93–96; David Childs, *The GDR: Moscow's Germany Ally*, 2nd ed. (London: Allen & Unwin Australia, 1988), 86.

8 For GDR efforts to shore up domestic tourism, see Moranda, *The People's Own Landscape*; for the promotion of East Berlin, see Michelle Standley, "The Cold War, Mass Tourism, and the Drive to Meet World Standards at East Berlin's T.V. Tower Information Center," in *Touring Beyond the Nation: The Development of Modern Tourism in a Pan-European and Transnational Context*, ed. Eric E. G. Zuelow (New York: Ashgate, 2011), 215–39; and Standley, "The Cold War Traveler."

9 David Childs, "The SED Faces the Challenges of *Ostpolitik* and *Glasnost*," in *East Germany in Comparative Perspective*, ed. David Childs, Thomas A. Baylis, and Marilyn Rueschemeyer (London: Routledge, 1989), 5.

10 For Cold War tourism as a historical travel culture and the West Berlin's travel organization's ambivalence, see Michelle Standley, "From Bulwark of Freedom to Cosmopolitan Cocktails: The Cold War, Mass Tourism, and the Marketing of West Berlin

as a Tourist Destination," in *Divided, But Not Disconnected: German Experiences of the Cold War*, ed. Tobias Hochscherf, Christoph Laucht, and Andrew Plowman (New York: Berghahn, 2011), 105–18.

11 "Einschätzung der Tourist-Saison 1970 in der Hauptstadt der DDR mit Schluß-folgerungen für die Vorbereitung der Saison 1971," 8 January 1971, C REP 100-05, Nr. 1454, Magistrat von Groß-Berlin, Abteilung Fremdenverkehr, Touristik und Berlin-Werbung, Landesarchiv Berlin, Germany [LAB].

12 "Bericht über die Tätigkeit des Bereiches Stadtrundfahrten im August 1978," September 1978, C REP. 123, Acc. 2401, Bd. 1, Reisebüro der DDR, Bezirksdirektion Berlin, Stadtrundfahrten, LAB.

13 For a time, the West Berlin-based coach tour company, Berolina, had the same name as East Berlin's official travel organization, which was first Berolina and then Berlin-Werbung Berolina, and finally in 1968 changed its name to the more international sounding, Berlin-Information. For the history of East Berlin's travel organizations see Standley, "The Cold War Traveler," 201–44.

14 In late 1971 the possibility of reorganizing responsibility for tourism in East Berlin was raised, because the city was struggling to cope with the sudden influx of foreign tourists. One proposal was that Berlin-Information assume full responsibility, but it did not have enough staff; another was to bring all tourism services in under the umbrella of Reisebüro-DDR Berlin. The outcome is not clear, but it seems likely that the latter option won out. "Vermerk über die Abteilungsleitersitzung am 17.12.71 der Berlin-Information," December 1971, C Rep. 123, Zg. 1707, Bd. 27, LAB.

15 "Bericht über die Tätigkeit des Bereiches Stadtrundfahrten im Juni 1977," July 1977, C REP. 123, Acc. 2401, Bd. 1, Reisebüro der DDR, Bezirksdirektion Berlin, Stadtrundfahrten, LAB.

16 Ellen Lentz, "What's Doing in Berlin," *New York Times*, 20 February 1977, section 10, p. 11.

17 "Bericht über die Tätigkeit des Bereiches Stadtrundfahrten im Januar 1978," February 1978, C REP. 123, Acc. 2401, Bd. 1, Reisebüro der DDR, Bezirksdirektion Berlin, Stadtrundfahrten, LAB.

18 "Zu politisch-ideologischer Fragen des Auslandstourismus der DDR," 13 October 1977, Rep. 200, 8.3.2, Nr. 5, Aufnehmende Auslandstourismus (AAT), Archiv Greifswald, Germany. "HOG" (Handels-Organisation Gaststätte) referred to Zenner's status as part of the Handelsorganisation (HO), the state-run trade organization.

19 Erik Cohen, "The Tourist Guide: The Origins, Structure and Dynamics of a Role," *Annals of Tourism Research* 12 (1985), 166; Christopher J. Holloway, "The Guided Tour: A Sociological Approach," *Annals of Tourism Research* 8, no. 3 (1981), 381–82; for an analysis of tourist and tour guide interaction, see Philip L. Pearce, "Tourist – Guide Interaction," *Annals of Tourism Research* 11 (1984), 129–46.

20 Surprisingly little has been written about bus tours. For an example of a recent study, see Tim Edensor and Julian Holloway, "Rhythmanalysing the Coach Tour: The Ring of Kerry, Ireland," *Transactions* (2008), 483–501.

21 Cohen, "The Tourist Guide," 22.

22 "Bericht über die Tätigkeit des Bereiches Stadtrundfahrten im März 1977," April 1977, C REP. 123, Acc. 2401, Bd. 1, Reisebüro der DDR, Bezirksdirektion Berlin, Stadtrundfahrten, LAB.

23 "Bericht über die Tätigkeit des Bereiches Stadtrundfahrten im April 1978," May 1978, C REP. 123, Acc. 2401, Bd. 1, Reisebüro der DDR, Bezirksdirektion Berlin, Stadtrundfahrten, LAB.

24 "Bericht über die Tätigkeit des Bereiches Stadtrundfahrten im September 1977," October 1977, C REP. 123, Acc. 2401, Bd. 1, Reisebüro der DDR, Bezirksdirektion Berlin, Stadtrundfahrten, LAB.

25 Ibid.

26 "Bericht über die Tätigkeit des Bereiches Stadtrundfahrten im October 1977," November 1977, C REP. 123, Acc. 2401, Bd. 1, Reisebüro der DDR, Bezirksdirektion Berlin, Stadtrundfahrten, LAB.

27 *Baedekers DDR* (Stuttgart: Karl Baedeker, 1989), 202. Still associated with the good times of their GDR past, Zenner continues to be a popular destination for older East Berliners, especially in the summer.

28 "Bericht über die Tätigkeit des Bereiches Stadtrundfahrten im Juli 1978," August 1978, C REP. 123, Acc. 2401, Bd. 1, Reisebüro der DDR, Bezirksdirektion Berlin, Stadtrundfahrten, LAB.

29 Ibid.

30 "Bericht über die Tätigkeit des Bereiches Stadtrundfahrten im Juni 1978," July 1978, C REP. 123, Acc. 2401, Bd. 1, Reisebüro der DDR, Bezirksdirektion Berlin, Stadtrundfahrten, LAB.

31 Ibid.

32 "Bericht über die Tätigkeit des Bereiches Stadtrundfahrten im Juli 1978," August 1978, C REP. 123, Acc. 2401, Bd. 1, Reisebüro der DDR, Bezirksdirektion Berlin, Stadtrundfahrten, LAB.

33 "Bericht über die Tätigkeit des Bereiches Stadtrundfahrten im Januar 1978," February 1978, C REP. 123, Acc. 2401, Bd. 1, Reisebüro der DDR, Bezirksdirektion Berlin, Stadtrundfahrten, LAB.

34 "Bericht über die Tätigkeit des Bereiches Stadtrundfahrten im März 1978," April 1978, C REP. 123, Acc. 2401, Bd. 1, Reisebüro der DDR, Bezirksdirektion Berlin, Stadtrundfahrten, LAB.

35 "Bericht über die Tätigkeit des Bereiches Stadtrundfahrten im September 1977," October 1977, C REP. 123, Acc. 2401, Bd. 1, Reisebüro der DDR, Bezirksdirektion Berlin, Stadtrundfahrten, LAB.

36 "Bericht über die Tätigkeit des Bereiches Stadtrundfahrten im Juli 1978," August 1978, C REP. 123, Acc. 2401, Bd. 1, Reisebüro der DDR, Bezirksdirektion Berlin, Stadtrundfahrten, LAB.

37 "Bericht über die Tätigkeit des Bereiches Stadtrundfahrten im März 1978," April 1978, C REP. 123, Acc. 2401, Bd. 1, Reisebüro der DDR, Bezirksdirektion Berlin, Stadtrundfahrten, LAB.

38 "Bericht über die Tätigkeit des Bereiches Stadtrundfahrten im Januar 1978," February 1978, C REP. 123, Acc. 2401, Bd. 1, Reisebüro der DDR, Bezirksdirektion Berlin, Stadtrundfahrten, LAB.

39 "Bericht über die Tätigkeit des Bereiches Stadtrundfahrten im März 1978," April 1978, C REP. 123, Acc. 2401, Bd. 1, Reisebüro der DDR, Bezirksdirektion Berlin, Stadtrundfahrten, LAB.

40 "Bericht über die Tätigkeit des Bereiches Stadtrundfahrten im April 1978," May 1978, C REP. 123, Acc. 2401, Bd. 1, Reisebüro der DDR, Bezirksdirektion Berlin, Stadtrundfahrten, LAB.

41 Ibid.

42 "Bericht über die Tätigkeit des Bereiches Stadtrundfahrten im Januar 1978," February 1978, C REP. 123, Acc. 2401, Bd. 1, Reisebüro der DDR, Bezirksdirektion Berlin, Stadtrundfahrten, LAB.

43 "Bericht über die Tätigkeit des Bereiches Stadtrundfahrten im Juni 1978," July 1978, C REP. 123, Acc. 2401, Bd. 1, Reisebüro der DDR, Bezirksdirektion Berlin, Stadtrundfahrten, LAB.

44 Gilbert Rist argues that Truman's fourth point in his speech to the UN on 20 January 1949 marked a paradigm shift, ushering in the "age of development." In Truman's definition, global development was a shared aim – regardless of which side of the ideological divide one was on – that could only be brought about by an influx of foreign aid and modern technology, and measured by the amount of goods one produced and amassed. See Gilbert Rist, *The History of Development: From Western Origins to Global Faith* (London: Zed, 2002).

45 Though derided by some commentators in West Germany and the US as a Disneyfied version of the past, both sites were enormously popular with East Germans and foreign

visitors, suggesting that in this case the party-state authorities of East Berlin were perhaps more alert to the zeitgeist and a general longing for a sanitized, romantic version of the premodern, pre-war German past.

46 "Bericht über die Tätigkeit des Bereiches Stadtrundfahrten im Juli 1979," August 1979, C REP. 123, Acc. 2401, Bd. 1, Reisebüro der DDR, Bezirksdirektion Berlin, Stadtrundfahrten, LAB.
47 "Bericht über die Tätigkeit des Bereiches Stadtrundfahrten im August 1977," September 1977, C REP. 123, Acc. 2401, Bd. 1, Reisebüro der DDR, Bezirksdirektion Berlin, Stadtrundfahrten, LAB.
48 "Bericht über die Tätigkeit des Bereiches Stadtrundfahrten im April 1978," May 1978, C REP. 123, Acc. 2401, Bd. 1, Reisebüro der DDR, Bezirksdirektion Berlin, Stadtrundfahrten, LAB.
49 "Bericht über die Tätigkeit des Bereiches Stadtrundfahrten im Oktober 1977," November 1977, LAB, C REP. 123, Acc. 2401, Bd. 1, Reisebüro der DDR, Bezirksdirektion Berlin, Stadtrundfahrten, LAB.
50 "Bericht über die Tätigkeit des Bereiches Stadtrundfahrten im September 1977," October 1977, C REP. 123, Acc. 2401, Bd. 1, Reisebüro der DDR, Bezirksdirektion Berlin, Stadtrundfahrten, LAB.
51 "Bericht über die Tätigkeit des Bereiches Stadtrundfahrten im März 1977," April 1977, C REP. 123, Acc. 2401, Bd. 1, Reisebüro der DDR, Bezirksdirektion Berlin, Stadtrundfahrten, LAB.

References

Baedekers DDR. Stuttgart: Karl Baedeker, 1989.
Betts, Paul, and Katherine Pence, eds. *Socialist Modernity: East German Everyday and Cultural Politics*. Ann Arbor: University of Michigan Press, 2008.
Childs, David. *The GDR: Moscow's Germany Ally*. 2nd ed. London: Allen & Unwin Australia, 1988.
———. "The SED Faces the Challenges of *Ostpolitik* and *Glasnost*." In *East Germany in Comparative Perspective*, edited by David Childs, Thomas A. Baylis and Marilyn Rueschemeyer, 1–18. London: Routledge, 1989.
Cohen, Erik. "The Tourist Guide: The Origins, Structure and Dynamics of a Role." *Annals of Tourism Research* 12 (1985): 5–29.
Colomb, Claire. *Staging the New Berlin: Place Marketing and the Politics of Urban Reinvention Post-1989*. London: Routledge, 2011.
Crew, David F. *Consuming Germany in the Cold War*. Oxford: Berg, 2003.
Edensor, Tim and Julian Holloway, "Rhythmanalysing the Coach Tour: The Ring of Kerry, Ireland." *Transactions* (2008): 483–501.
Endy, Christopher. *Cold War Holidays: American Tourism in France*. Chapel Hill: University of North Carolina Press, 2004.
Gorsuch, Anne E., and Diane P. Koenker, eds. *Turizm: The Russian and East European Tourist Under Capitalism and Socialism*. Ithaca: Cornell University Press, 2006.
Hixson, Walter L. *Parting the Curtain: Propaganda, Culture, and the Cold War: 1945–1961*. New York: St Martin's Press, 1997.
Holloway, Christopher J. "The Guided Tour: A Sociological Approach." *Annals of Tourism Research* 8, no. 3 (1981): 377–402.
Koshar, Rudy. *German Travel Cultures*. New York: Berg, 2000.
Lanfant, Marie-Françoise. "International Tourism, Internationalization and the Challenge to Identity." In *International Tourism: Identity and Change*, edited by Marie-Françoise Lanfant, John B. Allcock, and Edward M. Bruner, 24–43. London: SAGE, 1995.

Lentz, Ellen. "What's Doing in Berlin." *New York Times*, 20 February 1977, section 10, p. 11.

Mitter, Rana, and Patrick Major, eds. *Across the Blocs: Cold War Social and Cultural History*. London: Frank Cass, 2004.

Moranda, Scott. *The People's Own Landscape: Nature, Tourism, and Dictatorship in East Germany*. Ann Arbor: University of Michigan Press, 2014.

Pagenstecher, Cord. *Der Bundesdeutsche Tourismus: Ansätze zu einer Visual History: Urlaubsprospekte, Reiseführer, Fotoalben 1950–1990*. Hamburg: Kovac, 2003.

Pearce, Philip L. "Tourist – Guide Interaction." *Annals of Tourism Research* 11 (1984): 129–46.

Poiger, Uta. *Jazz, Rock, Rebels: Cold War Politics and American Culture in Divided Germany*. Berkeley & Los Angeles: University of California Press, 2000.

Rist, Gilbert. *The History of Development: From Western Origins to Global Faith*. London: Zed, 2002.

Sarotte, M. E. *Dealing with the Devil: East Germany, Détente and Ostpolitik, 1969–1973*. Chapel Hill: University of North Carolina Press, 2001.

Spode, Hasso, ed. *Goldstrand und Teutonengrill. Kultur- und Sozialgeschichte des Tourismus in Deutschland 1945–1989*. Berlin: W. Moser Verlag für Universitäre Kommunikation, 1996.

———. *Zur Sonne, zur Freiheit! Beiträge zur Tourismusgeschichte*. Berlin: W. Moser Verlag für Universitäre Kommunikation, 1991.

Standley, Michelle. "From Bulwark of Freedom to Cosmopolitan Cocktails: The Cold War, Mass Tourism, and the Marketing of West Berlin as a Tourist Destination." In *Divided, But Not Disconnected: German Experiences of the Cold War*, edited by Tobias Hochscherf, Christoph Laucht, and Andrew Plowman, 105–118. New York: Berghahn, 2011.

———. "The Cold War, Mass Tourism, and the Drive to Meet World Standards at East Berlin's T.V. Tower Information Center." In *Touring Beyond the Nation: The Development of Modern Tourism in a Pan-European and Transnational Context*, edited by Eric E. G. Zuelow, 215–39. New York: Ashgate, 2011.

———. "The Cold War Traveler: Mass Tourism in Divided Berlin, 1945–1979." PhD diss., New York University, 2011.

———. "'Here Beats the Heart of the Young Socialist State': 1970s East Berlin as Socialist Bloc Tourist Destination." *Journal of Architecture* 18, no. 5 (2013): 683–98.

Whitfield, Stephen J. *The Culture of the Cold War*. Baltimore: Johns Hopkins University Press, 1991.

Part II
Encounters

4 The Artek camp for Young Pioneers and the many faces of socialist internationalism

Kathleen Beger

Shortly after its foundation, Artek, the holiday camp for Young Pioneers, was made the flagship for the Soviet Union's achievements in the fields of child welfare, pedagogy, and internationalism, celebrated in an enormous amount of literature and visual media. A passage taken from a photo book published in 1976 recorded the impressions of a young Soviet girl who had stayed there:

> I am thankful to Artek for everything. Here, I truly understood the role our country plays in the world. Here, I got to know children from Vietnam. In their lands bombs are exploding, schools are burning. All honest people help Vietnam. Our Fatherland always stands for peace. And for us, guys, it is clear that children want peace, friendship for the entire world. Nadia Plekhanova, Chelyabinskaya oblast.[1]

Although the authenticity of this statement is questionable, it tells us what image Artek was supposed to communicate to both its domestic and foreign audiences. The message was fairly obvious: the Soviet Union wanted to be perceived as a country fighting for peace, international solidarity, and friendship.

This chapter seeks to shed light on Artek's multinational character and the many facets of its socialist internationalism, focusing on the period between the late 1950s and 1980s. It describes the strains that arose from the camp visitors' multi-ethnic and multinational composition, and compares different perspectives: those of the children and adolescents from Soviet, socialist, and capitalist backgrounds as well as those of the camp's Pioneer leaders and directors.

Founded by the Russian Red Cross Society in 1925, Artek was initially intended for working- and peasant-class children suffering from tuberculosis.[2] Due to its innovative combination of recreational and educational measures and its picturesque setting on Crimea's Black Sea coast, it became a model for other Pioneer camps.[3] As Artek grew and became better equipped over the course of its first decade, Soviet Pioneers were soon sent there based on their behaviour and good contacts rather than their health. From the mid-1930s onwards, a holiday at Artek was considered an honour, rewarding heroic deeds, excellent school results, and Pioneer achievements.[4] The camp's increasing prestige made it a popular summer destination for the party elite's offspring as well.[5]

However, Artek did not only welcome Soviet Pioneers. Serving as a political showcase for the promotion of communism and demonstrating its superiority over capitalism from the very beginning, its leadership also invited foreign children's delegations.[6] This function took on a new significance after Nikita Khrushchev came to power. The de-Stalinisation campaigns put an end to the country's isolationism, which had begun in the second half of the 1930s and reached its climax during late Stalinism. Calling for "peaceful coexistence" at the 20th Communist Party Congress in 1956, Khrushchev paved the way for a new level of international relations with both socialist and capitalist countries. He emphasised the importance of mutual trust, development, and support and revived the model of internationalism favoured in the 1920s and early 1930s.[7]

Soviet citizens were now allowed to travel abroad, yet though numerous cultural exchange agreements were signed with socialist and capitalist countries, it did not mean that citizens could go abroad easily. Rather the opposite.[8] Khrushchev's primary intention for raising the Iron Curtain was not so much to move people, but to enable the transfer and exchange of professional knowledge and, to some degree, culture.[9] Informal encounters between Soviet citizens and foreigners were anything but welcome, and the chances of coming across a "real" foreigner remained extremely unlikely for the vast majority of the country's population.

Under these circumstances, Artek can be considered a special place, which benefited exceptionally from Khrushchev's turn in foreign policy. From 1957 onwards, the camp held yearly "international summer weeks" to which it invited children's and youth groups from socialist and capitalist countries. In 1966, for example, the camp was visited by 365 children and adolescents plus 40 delegation leaders from 34 capitalist countries, as well as 359 Pioneers and 27 Pioneer leaders from ten socialist countries.[10] During their stay, they were obliged to take part in events designed to foster friendship, solidarity, and peace. Over time, the international summer weeks became something of an institution, following a highly ritualised pattern. To present the Soviet state and Artek in the best light and to reduce the risk of importing "Trojan horses," the activities during these weeks were meticulously organised to keep the camp's internal order under control and to prevent undesired outcomes.[11] The international summer weeks are best understood as examples of what Erving Goffman has called "front region" and "front-stage behavior."[12] Drawing on Dean MacCannell, who refined Goffman's concept, I will demonstrate that Artek's temporary inhabitants, Soviet and foreign, attached far more importance to the unofficial, "backstage," or "off-stage" get-togethers than to any staged celebration of international friendship, solidarity, and peace.[13] Consequently, they found ways to establish close, spontaneous contacts with their peers from all over the world. As their ambitions and interests often ran counter to the camp's official objectives, Artek faced challenging situations. In trying the solve them, the camp's leadership often came up with measures that were contradictory and ineffective, with at times paradoxical results.

I have based my discussion on the records held in the Russian State Archive of Socio-Political History (RGASPI), the Russian State Archive (GARF), the Bundesarchiv in Berlin-Lichterfelde (SAPMO-BArch), primary (official) sources, and secondary literature. Moreover, I use seven biographical interviews, which I conducted in Russia and Germany between 2015 and 2017.[14] Four of the respondents come from the former Soviet Union, two from the Federal Republic of Germany (FRG), and one from the former German Democratic Republic (GDR). As there is no archival material on Artek for the late 1970s onwards, and since such archival material as survives provides little insight into how individual visitors experienced staying at the camp, the primary purpose of these interviews was to generate additional historical data.[15]

Staging socialist internationalism

Immediately after the October Revolution, internationalism became one of the main ideological mainstays of the Soviet Union. To promote communism, the Soviet state maintained close contacts with foreign socialist and communist parties.[16] Soviet children were educated in the spirit of internationalism and urged to "stand up for the cause of the working class in its struggle for the liberation of the workers and peasants of the entire world."[17] Thus, in the 1920s and early 1930s, the country's Pioneers were expected to assist their peers abroad in their fight for communism by holding parades and demonstrations, or initiating charitable collections.[18] However, when it became apparent that the expected class wars would not occur, Stalin abandoned the idea of the "world revolution," and replaced it with the theory of "socialism in one country."[19] Since the early 1930s, internationalism noticeably merged with Soviet patriotism in the shape of the so-called "druzhba narodov," the friendship among Soviet and later socialist nations.[20] After Stalin's death in 1953, however, Khrushchev re-introduced the model of "*proletarian* (in relation to international communist parties in power) and *socialist* internationalism (towards the rest of the world)" but combined these concepts with a new cultural internationalism.[21]

The children's internationalism of the 1920s and early 1930s was directed at the promotion of communism. In contrast, post-war internationalism was predominantly concentrated on the demand for peace.[22] The Great Soviet Encyclopaedia of 1972 defined "socialist internationalism" as a "new type of international relations, formed and developed on the basis of friendship, equality, mutual respect, . . . fraternal cooperation, political, economic, military and cultural mutual support of nations and nationalities who embarked upon the path of socialism."[23]

During the Thaw, the Soviet concept of the enemy underwent a significant change too. Ted Hopf has argued that "The principle of 'who is not against us is potentially with us' replaced the Stalinist principle of 'who is not with us is against us,'" intensifying above all Soviet relationships with countries in Asia, Africa, and Latin America, and later on with Western states.[24] Building on this, Pia Koivunen has shown that "once clear and strict boundaries between friends and enemies became blurry, and instead of the old binary model a new category emerged: 'a potential friend.'"[25] This new approach in foreign policy was reflected in the set-up of Artek's international summer weeks, as the Komsomol invited children and youth delegations consisting of young people aged 12 to

15 years, both from socialist and capitalist countries. Those from Western states, however, were always delegates of socialist or communist youth organisations, which meant that the plan was for international friendship to be celebrated with like-minded foreigners.[26]

The daily schedule of the international summer weeks barely varied before the late 1980s. Each year, all delegations were expected to contribute to the so-called "days of international friendship" and "national days," to participate in political discussions and camp-wide sports competitions. When sending out invitations to foreign organisations, the Komsomol also informed them about the international summer weeks' programme and activities so that they could prepare their members for their stay.[27] Usually, these weeks either carried a specific motto or commemorated an anniversary of great significance. Those of 1969 and 1976 were dedicated to Lenin's 100th birthday and the 60th anniversary of the October Revolution respectively, while the weeks of 1977 focused on the theme "For a happy childhood in a world of peace."[28]

According to three of my interviewees, one from the GDR and two from the FRG, the children and adolescents in their delegations took their stay at Artek seriously, performing preliminary activities and informative discussions several months before their departure.[29] It was particularly important to prepare for the celebration of various national days, which provided each delegation with the opportunity to put on an exhibition about its homeland. Such exhibitions provided not only general information, but also presented traditions, folklore, music, food or aspects of daily life, leisure, and hobbies. For this purpose, the delegations' members started to collect specific information, photographs or newspaper articles, and took them to Artek. The official reason for these exhibitions was to impart knowledge about different countries to the camp's young visitors.[30]

During the days of international friendship, all delegations came together in Artek's stadium, where one or two representatives of each group would give a speech about how they contributed to the preservation of peace in their home countries. In the 1960s and 1970s, special attention was given to Vietnam and Chile – countries which were living in a state of war and violence.[31] Some of the delivered speeches may have been quite empathic and emotional, but the days of international friendship as a whole were still highly ritualised. As Alexei Yurchak has demonstrated, the Soviet discourse of public ritual, in general, "became increasingly standardized, following a unified and centrally orchestrated scheme" in the late 1950s.[32] Once "designed and conducted by local social, cultural, and educational institutions," civic rituals, Yurchak argues, "became united into one centralized 'system of rituals' run by the party."[33] The same held true for Artek's international summer weeks whose programme, compiled by Komsomol officials in Moscow together with the camp's local leadership in Crimea, consisted of well-planned and standardised events which essentially did not change until the late 1980s.

In this context, language played a crucial role. Finding a common understanding proved difficult as not all foreign delegations spoke Russian. Upon arrival at the camp, each foreign group was assigned one or two Soviet interpreters as well as Soviet Pioneer leaders. Despite this, understanding one another was still challenging. A GDR's delegation leader, who had accompanied a group of Pioneers to Artek in summer of 1963, remarked in his final report that some of the planned activities could not be attempted because of communication problems. For instance, the camp's international children's council consisting of young representatives from each foreign and Soviet delegation was not always able to work independently – meaning without the help of adult supervisors.[34]

Except for the days of international friendship, political discussions on a smaller scale in which only a few delegations participated were organised as "friendship meetings." In the summer of 1963, a group of GDR Pioneers came together with peers from the FRG to exchange their views regarding the issue of German reunification. As the GDR's delegation leader noted in his report, his Pioneers were not able to answer the question "correctly." Hence he felt the need to instruct them several times so that they could "adequately" take part in the discussion. Highlighting this incident, he called for a more careful selection of potential Artek candidates. According to him, some of his delegation members were not worthy to travel to the Soviet Union's most prestigious Pioneer camp, either because they were unable to take firm political–ideological positions or because they were physically too weak to bear the hardships of the journey.[35] His summary reveals at least two important features. Firstly, in some cases, the Soviet leadership of Artek was less concerned with controlling international gatherings as one might expect. Secondly, not only the Soviet Union, but also other socialist states used Artek's international summer weeks to present themselves in the best light and to win recognition.

Intercultural clashes

Even the most careful preparation could not completely eliminate undesired and unintended outcomes. International gatherings always bore the risk of confrontation. The sheer number of young people from various backgrounds, traditions, and cultural habits gathering in the camp inevitably led to tension. A variety of misunderstandings between Soviet hosts and their guests occurred during the very first two international summer weeks of 1956 and 1957, as a report written by the camp's deputy director revealed. The document provides a fascinating insight into the camp's multinational character, because it depicted each foreign group and subjectively listed both positive and negative aspects of each delegation's behaviour and demeanour.[36]

The author focused on the behaviour of the children and their group leaders. For instance, the Belgian delegation leader was criticised for her loving interaction with children. As Soviet pedagogy rested upon strictness and discipline, kissing children and tenderly stroking the children's heads was met with incomprehension by Artek's deputy director.[37] He also disapproved the pedagogy of Yugoslavia, the FRG, Austria, and Sweden for leaving all the decisions up to the children themselves. Referring to the Yugoslav delegation, his report remarked that

> The group leaders' educational principles . . . lead to total 'freedom', i.e. in our view to the complete anarchy of the child's soul. . . . For 45 minutes they

did not come out of the water, wore dirty training clothes, reached poor results in sports competitions, did not know a lot of songs etc. . . . . In all discussions they expressed anguish concerning the Yugoslav–Soviet Split of 1948, spoke evil of Stalin, exorbitantly praised Tito, considered him a 'second Lenin.'[38]

From the author's Soviet perspective, this "pedagogical freedom" did not bring positive results. Instead, it led to bad behaviour and poor physical and mental skills. The pedagogical frames of reference in the Khrushchev era, based on nineteenth-century and a few early-Soviet pedagogical theorists, were only "lightly reversed versions of Stalinist originals."[39] Soviet pedagogy remained teacher-centred and expected students to handle an excessive workload.[40] Children and adolescents were expected to be disciplined and to subordinate their own interests to those of the collective.[41] Moreover, it was politically sensitive to praise Tito as it shook the foundations of Soviet ideology, according to which Tito could never be regarded a "second Lenin." Although Khrushchev had discredited Stalin about six months before the arrival of the Yugoslavian group in Artek, criticising the Soviet system or Stalin himself was still unacceptable.

Interestingly, the reports of 1956 and 1957 directed criticism at Soviet Pioneers too. The camp's deputy directory was ashamed to hear that they had behaved arrogantly and disrespectfully towards their Hungarian peers. The latter were insulted by some Soviet Pioneers' assertion that Hungary was unable to exist without Soviet support, and that Soviet soccer players were much better than Hungarian ones.[42] At first glance, statements such as these seem to be typically childish confrontations. Yet, they were also a product of a patriotic upbringing, which had become an important part of Soviet pedagogy. Children were supposed to love their country, to be proud of its achievements, to value its efforts in terms of child welfare, and to understand how fortunate they were.[43]

There were also misunderstandings during the numerous competitions that were an integral part of Artek's daily activities. Competition was not only a driving force in capitalist societies; it existed under socialism as well, even though it possessed other forms and obeyed a different logic. By declaring that "competition would allow workers to 'display their abilities, develop capacities, and reveal those talents' suppressed under capitalism," Lenin had hoped to motivate workers to enthusiastically do their jobs without any means of control.[44] Competition was supposed to increase "both economic production and productivity" and, by doing so, to build "the ideal communist society."[45] In contrast to its capitalist counterpart, socialist competition was regarded "pure and unselfish."[46] Not individual and egoistic, though; only those interests that benefited society as a whole were to be privileged.[47]

At Artek, competitions between different delegations were held at all levels. Basically, any aspect of daily life could be turned into a contest, measured, and then used to evaluate the groups' behaviour and achievements. Emerging victorious from these competitions was the main reason why it was almost only Soviet "shock" Pioneers – those who attained the best grades and excelled in all areas – who were allowed to visit the camp during the summer months. The outcome was that Artek's international summer weeks were increasingly a stage for Soviet self-representation, as was noted by the other socialist countries. A meeting of the GDR's Ernst Thälmann Pioneer Organization secretariat concluded that not all the children it had sent to Artek in the summer of 1963 had met the camp's requirements.[48] Five years later, this problem was still an issue. According to one German Pioneer leader, it was not possible anymore to win a contest with "rank and file" Pioneers as they could not keep up with the competition. He recommended sending those children and adolescents who had already won local, regional, or national contests.[49] Thus, to present themselves in the best way possible and to impress the "big brother," Pioneers from other socialist countries had to pass through a strict selection process similar to that of their Soviet peers.

By contrast, self-presentation was less important to delegations from capitalist countries. While socialist groups were considered official representatives of their states, capitalist ones usually did not have this status or the responsibility that came with it. Western youth organisations did belong to a specific political party, and hence did not feel obliged to act in the name of the whole country. The Socialist Youth of Germany – The Falcons – from West Germany, for instance, was an independent youth organisation, although it temporarily shared political views with the Social Democratic Party (SPD). Peter, a former Falcons' leader who had accompanied a group to Artek in 1977, told me that his organisation had a different attitude towards the selection process of potential Artek candidates. As the Falcons consisted of stable local groups distributed over the whole country, each year one of these groups was chosen to travel to the Soviet Union, regardless of the group members' performance at school, in sports or the arts. Peter's group consisted of five boys and five girls from Gelsenkirchen. Most of them attended secondary schools; one boy went to a special needs school. None of them had ever been abroad before. Sure enough, during the competitions there was an imbalance between the Soviet and other socialist delegations and their capitalist peers, as Peter said:

> In Artek, we came together only with the country's children's elite. Thus, our two worlds clashed. . . . Once, M. [in Peter's group] had to participate in a quiz but she did not know who Rembrandt was. Most likely, the interpreter . . . did such a good job so that nobody noticed that she actually did not know the answer. Hence, she won the first prize.[50]

Due to this disparity, the competitions at Artek were mostly fought out between socialist states.[51] In order not to annoy some of their foreign guests, the Soviet personnel sometimes had to take extraordinary measures. However, Peter was generally astonished by the fact that "there was nothing that one does just because it is fun. Everything was embedded into competitions."[52]

Indeed, the peculiar Soviet understanding of (youth) tourism as a rational and purposeful leisure activity seemed somewhat strange to foreigners, especially those from capitalist countries. Holidaymakers, no matter whether children, adolescents or adults, were expected to further their personal development, to extend their knowledge, and to gain new strengths so that when returned to school or work, they would be able to work even more productively to build communism.[53] In this sense, competitions were not only a means for self-improvement but also for engaging in socially useful work. It was the well-being of society that was paramount, not individual pleasure.

That said, Artek's guests usually found ways to evade the official requirements. It seemed that, unlike their socialist peers, the children from capitalist countries did not feel the need to hide their incomprehension and simply refused to join in Artek's countless competitions. Monika, a young member of the Falcons who visited Artek in 1978, emphasised that her group always received positive or negative credits for certain behaviour or non-behaviour, for instance, for cleaning their rooms, marching in formation to the dining hall, or for singing. Monika's Soviet Pioneer leader insisted that they complete several competitive tasks every day. The adolescents mostly ignored her rules:

> It annoyed us. I assume that she [the Soviet Pioneer leader] was criticized because her delegation did what it wanted. But then . . . we decided we will do this at least once, just to show that we are able to do it. . . . The beds were made up, everything was clean and tidy, we were singing and marching . . . and won the challenge cup. The next day, we did as before. For us, it was important to demonstrate that we can if we want, but we do not like this.[54]

If nothing else, following all Artek's guidelines for the international summer weeks was a tough job when the group consisted predominantly of children and adolescents from capitalist states. Soviet Pioneer leaders had to learn to compromise if they wanted to create a pleasant atmosphere for all.

Another problematic topic was the (foreign) girls' interest in fashion and cosmetics, something that Komsomol officials often tried to supress.[55] In the Soviet Union, the question of appearance had already been widely discussed during the period of the New Economic Policy (NEP) in the 1920s, when political leaders praised fashion asceticism and condemned all signs of a "bourgeois look" such as high heels. Young Komsomol activists were required to take the working class as their example in all spheres of life, including fashion. A proper, modest outfit or look was supposed to emphasise their belonging to the state, which claimed to be ruled by the working class.[56] However, even under Stalin this ideal was fairly ambivalent as "the Soviet person was encouraged to enjoy consumption of personal 'bourgeois' pleasures (dresses, wristwatches, lipstick) as long as they were not used for egoistic goals of social prestige, careerism, and so forth, but as elements of 'cultural life' and due rewards for hard work."[57] In the post-war years, when Soviet youth began to listen to foreign radio channels and were more and more interested in Western dance, film, music and fashion, the question resurfaced as a political issue.[58] The Soviet regime's paradoxical, paranoid attitude towards Western influences made it difficult to distinguish between what was desirable and what was not.[59]

At Artek, conflicts of this kind occurred constantly, and were primarily caused by the fact that everyone – Soviet and foreign children and group leaders – was expected to wear a uniform. Immediately on arrival they had to hand over their own clothes in exchange for a uniform of shirts and trousers or skirts (a procedure which dated back to 1925). As appearance and individuality are typically important for adolescents, the Artek uniform faced stiff resistance from all quarters, whether Soviet, socialist, or capitalist. Gisela, who accompanied a group from the GDR to Artek in 1976 as a group leader, confirmed this:

> Our girls had nail varnish. This had to be removed! [The nails] were cut. . . . We had to hand over our suitcases and received a uniform. We took only personal things and underwear with us. . . . Our children did not really like this procedure, always the same clothes. . . . But after a while, they got used to it.[60]

According to a report by one GDR delegation leader who visited Artek in the summer of 1968, the camp organised a beauty contest that year. The first prize was awarded to a girl in his group.[61] While it is unclear whether competitions of this kind took place in subsequent years or not, the beauty contest of 1968 – and any later ones – seems to have been embedded in Artek's culture of competitions, and took place under the Soviet Pioneer leaders' surveillance and guidance.

In a study on former Soviet Pioneers who visited Artek in the late 1970s or 1980s, Anna Kozlova's interviewees remembered the uniforms as "fashionable, beautiful, and comfortable."[62] Yet Inna, one of my respondents from the former Soviet Union, who was at the camp in June 1977, brought up an interesting point. She suspected that in order to window dress the camp, it was equipped with new and better clothing for the international summer weeks only. As Inna stayed there just before the foreign groups arrived, she remembered wearing second-hand uniforms.[63] Boris, another Soviet Pioneer who was at Artek in the autumn of 1986 or 1987, hated his uniform and said it was "not fitting, worn, old and dirty."[64] In contrast, Monika was very impressed by the turquoise anoraks, to the point where she regretted her group's refusal to wear the Artek uniforms.[65]

While Soviet Pioneers and those from other socialist states had to wear the uniform, delegations from capitalist states had some leeway. The preferences of capitalist groups or their leaders determined whether they would wear uniform or not. As Grigorii, a former Soviet Pioneer leader and director of Artek's Diamond sub-camp, put it:

> The most important thing was to not forbid anything . . . I told each of the group leaders: 'We don't want to deprive you of your personality. We just want you to feel comfortable because the uniforms will be changed. Otherwise, you must wash them yourself, and this is difficult.' . . . After two, three days their personal clothes were dwindling, nobody wanted to wash, and, finally, all of them started to wear the uniform. . . . My predecessors said:

'Under no circumstances, this is a rule!' But I said: 'If you want you can wear [your own clothes],' because I knew that this would stop after a while.[66]

However, Grigorii's method was not always successful. Peter's and Monika's groups wore their own clothes throughout, for example.[67] As an article in the German newspaper *Stuttgarter Zeitung* illustrates, it was not only the Falcons, but also delegations from other Western countries that were unwilling:

> For foreign children, the intrusive and military-like ceremonial of the 'Young Pioneers' with goose step, roll calls and salutes is stupid. Those from Gelsenkirchen – like other Westerners – who kept aloof from this refused to wear the uniform shirts and were met with understanding from the Soviet side.[68]

The uniform was a recurring issue. Delegations from capitalist countries saw it as a symbol of militarism and oppression, designed to make individuals compliant, to indoctrinate them, to subsume them into the masses, and deprive them of their freedom. Groups from socialist states were generally less likely to take offence, as many were already used to it because of similar practices in their home countries. Beyond that, uniforms were deemed a legitimate means to blur the differences between children and adolescents from various backgrounds. For the camp's leadership, this was an important point. People from all over the world visited Artek, and the Komsomol officials wanted to ensure some degree of equality. There was a practical consideration too: the camp's architecture was designed for collective life, and dormitories were so small that there was neither place for privacy nor storage for personal belongings.[69]

Benefitting from internationalism

The regulations notwithstanding, Artek was still a rather liberal Pioneer camp. Similar Soviet institutions elsewhere operated along far stricter lines. As Peter said, a group of the Falcons found this out when they stayed at the Orlenok camp on the Black Sea coast in the region of Krasnodar. Instead of contact with Soviet Pioneers, they were only allowed to form ties with another Western delegation from France.[70] At Artek, however, the children from capitalist states were usually not separated from those from socialist ones. They were assigned to the same organisational units (called *otriad*) and shared dormitories. Monika's *otriad*, for instance, consisted of children from the FRG, the Soviet Union, the US, and South Africa. She remembers falling in love with a Soviet Pioneer, Igor, at first. But after the arrival of the American group, Oscar from New York won her affections.[71]

As travelling abroad was out of reach for the vast majority of Soviet Pioneers, close contact with people their own age from all over the world made a lasting impression. Svetlana, who attended the international summer weeks in 1982, still vividly remembers one boy from Japan:

> He had invited me . . . we sat down, observed the stars and talked to each other. He spoke Japanese, I spoke Russian. It is difficult to say that we understood

each other, but the presents he gave to me, I keep until today. At home I have a small box. . . . Today it is funny. There are small coloured balls made of glass and . . . cranes made of paper.[72]

Although a Soviet Pioneer's stay in the camp was supposed to be a one-time occurrence, some managed to return as professionals. One of them was Grigorii who had first been there as a teenage boy in spring of 1974. Overwhelmed by the atmosphere, he had sworn he would come back. After finishing school, he began to study English language and literature. One day, he decided to take his destiny into his own hands, travelled to Crimea, introduced himself to Artek's leadership, and was eventually offered a job as a student interpreter for the international summer weeks. Starting in the early 1980s, he first worked as a Pioneer leader and later as the director of the Diamond sub-camp. For him and other language students, Artek provided ideal conditions for interacting with native speakers, and practising and improving their foreign-language skills. Grigorii benefited greatly from this unique privilege:

> In camp Diamond . . . I supervised 420 children and about 70 employees. . . . Here, I worked together with delegations from the USA, the GDR, Angola, Poland, Finland, Sweden, Algeria. I . . . really enjoyed the work during the international summer weeks, because I invited all foreign group leaders for a cup of tea to my office in the evening, where we had nice conversations. . . . My task was to guarantee good relationships between the Pioneer leaders, the foreign group leaders and the translators . . . In the Soviet Union, you could not go anywhere, it was difficult to make a tourist trip. That is why Artek was such a unique place. I could talk to everybody, to whom I wanted, my English skills helped me a lot and nobody controlled me regarding what I said to whom and why.[73]

Because he spoke English, Grigorii was able to have direct contact with the foreign delegations' leaders. In his own words, his work even gave him a sense of personal freedom. At a time when foreign travel was impossible for most Soviet citizens, Artek was a golden opportunity for a select few. However, as Anna Kozlova has shown, former Artek employees generally tend to understate or deny the camp's political-ideological character. Due to its exotic location on the Soviet periphery, staff members often distinguished it from the Soviet centre and its ideology, emphasising the experience of an "autonomous" childhood or job, and saying it was an " 'exception' from the usual Soviet everyday life."[74] Paradoxically, people like Grigorii, who first came as Pioneers and then returned as professionals, can be regarded the perfect "product" of Artek's mission: educating the Soviet Union's future political elite. Then again, his responsibilities notwithstanding, Grigorii never insisted on obeying the camp rules. He was fairly open-minded when it came to negotiations. This attitude was not unselfish. Although Grigorii had to take his foreign guests' wishes into account to guarantee a pleasant atmosphere in his camp, he was very interested in their well-being because it was the essential precondition for having contact with them.[75] By working at the camp from the late 1970s onwards, Grigorii built up a worldwide network of friends he had made at Artek. He was able to benefit from it years later during perestroika and after the Soviet Union's collapse. In 1990, he was invited to the US to speak to an association that organises worldwide youth camps. Thus, Grigorii was successful in transferring the cultural, social, and symbolic capital he had gained under Soviet conditions into post-Soviet life. Today he works for one of the world's leading associations for international youth exchanges. Grigorii's time at Artek as a young Pioneer turned out to be the first step in a successful international career.[76]

As the experiences of Grigorii, Svetlana, and Monika reveal, officially staged events at Artek did not hold their attention. What really mattered to them was the spontaneous, close contact with peers and colleagues from around the globe. At least, this is how they remember it when interviewed today. Due to its multinational composition and its beautiful subtropical setting, surrounded by the Black Sea coast and the Crimean Mountains, Artek was understood to be an exotic and extraordinary site, or, using a term coined by Anne E. Gorsuch, a "Soviet abroad."[77]

Limits of internationalism

Making new friends at Artek was one thing, keeping those ties alive after Artek was another. Love affairs, especially those between people from the Soviet Union and capitalist states, most clearly displayed the limits of international friendship. While at Artek, these relationships were widely accepted; the problems occurred after people returned home and tried to stay in touch. Oksana, who participated in the international summer weeks of 1961, faced such a situation:

> And there I met my first love, a boy from Austria, and, of course, after that [I had] a very bad feeling because we could not write to each other, i.e. at the beginning we wrote each other. But then, . . . my parents . . . said that I should better stop writing. But I remember his address to this day, in Vienna.[78]

Oksana explained that the reason she had to stop corresponding was not only because her parents were afraid of regularly receiving letters from a capitalist country. They were also concerned about their daughter's feelings and wanted to protect her from heartbreak. Oksana and her parents knew that, things being what they were, there was no chance she would see the Austrian boy again.[79]

Restrictions existed between socialist states too. While a student, Grigorii worked not only at Artek, but also at the Pioneer Republic Wilhelm Pieck near East Berlin, where he fell in love with a woman from the GDR:

> When I worked in the Pioneer Republic, I met a remarkable young woman. To leave the Soviet Union a special visa was required, . . . an "exit visa." But they did not issue me that visa. If they had given it to me, I would have had a German wife today.[80]

In general, Soviet rulers viewed "real" interpersonal relationships between people of different nationalities and across borders with scepticism. In fact, such relationships rarely went beyond staged and controlled events that are best described with Anne E. Gorsuch's phrase as " 'friendship' at a distance."[81] Furthermore, camp directors like Grigorii had to take into consideration their guests' foreign-policy interests in order to avoid conflict:

> Of course, . . . the leadership of Artek distributed the various foreign delegations with regard to the global political situation. The delegation from Israel in Camp Cypress wanted to be far away from groups coming from Arab countries so that their paths did not cross. . . . We also had a delegation from Algeria. Its leader was always against Israel, against the USA, against everything, even though, in our camp, Algeria, the USA and Israel lived far away from each other.[82]

Likewise, for delegations from the GDR and the FRG, coming into contact was difficult, if not impossible. Usually, encounters only took place during official, carefully staged meetings. Peter mentioned that his Falcons did not have a single encounter with their peers from the GDR at Artek, and while Monika recalled that she did indeed meet GDR Pioneers there, it was at an official get-together, accompanied by several group leaders.[83] The meeting left Monika with a strange feeling, as they had nothing but good to say about their lives in the GDR, where everything was said to be great, but she did not believe their exaggerated stories and reckoned the event was ideologically driven and rather counterproductive.

Conclusion

Khrushchev's change in foreign policy enabled the Artek Pioneer camp to organise yearly large-scale international summer weeks, in which children's and youth delegations from around the globe took part. Using the camp as a stage, the Soviet Union sought to improve its image, demonstrate its superiority, and impress its foreign guests. This was most evident on international friendship days, which were designed to promote solidarity and peace.

Implementing the Soviet branding presented a challenge to Soviet Pioneer leaders and camp directors, though, as they had to ensure the well-being of all domestic and foreign delegations and at the same time dispose of all undesired foreign influences. The discrepancy between Artek's official objectives and visitor expectations meant the camp staff ended up having to square the circle. Their attempts often resulted in misunderstandings, confrontations, and paradoxical situations – such as the official ban of cosmetics and the requirement to wear Artek uniform, followed by holding a beauty contest. Welcoming people of different national, cultural, and traditional backgrounds to Artek, the camp staff could not expect that all guests would follow the regulations to the letter. Living together at Artek entailed constant negotiation, and the inevitable redefinitions of its rules.

The camp's principles and official activities did not transform foreigners, especially those from capitalist states, into enthusiasts for socialism. That is not to say that they disliked their stay at Artek. But unlike the girl quoted at the very beginning of this chapter, visitors either remembered the official events such as the international friendship days as highly ritualised, stereotyped performances, or forgot them altogether. Artek's visitors were driven by a search for greater authenticity, for "real" face-to-face encounters with their peer group from abroad. This did not necessarily exclude the Soviet Pioneers and their leaders, or even the camp's directors. The Soviet Union's campaigns to promote friendly, solidary, and peaceful relations with people from other countries provided the ideological justification for such encounters. Over time, Soviet citizens began to appropriate the discourse of socialist internationalism, linking it to understandings and interpretations of their own. In an era when informal get-togethers with foreigners were unimaginable for Soviet citizens, Artek was a unique space where such encounters were possible. Soviet Pioneers, Pioneer leaders, and camp directors found ways of circumventing the official requirements, asserting their own interests by establishing personal, informal contacts with foreigners. However, Artek was a law unto itself in many ways. "Real" interpersonal relationships between Soviet and foreign citizens were unacceptable to the Soviet authorities, and many camp participants were forced to drop their contacts. As in other realms, the late Soviet system found itself struggling with the consequences of social practices that it had generated with its own propaganda. In this sense, the Artek camp for Young Pioneers was an impressive example of what Alexei Yurchak has observed, namely that Soviet discourses such as the discourse of internationalism, and Soviet rituals such as its uniforms, could create alternative spaces for personal freedom and cross-national encounters, engendered in the interaction between the young travellers and their Soviet hosts.

Notes

1 Evgenii Rybinskii, *Pionerskaia respublika Artek: Fotoal'bom* (Moscow: Planeta, 1976), 199.
2 In 1937, responsibility for administering the camp was transferred from the Russian Red Cross Society to the People's Commissariat of Health (Evelin Eichler, *Pionierlager in der Sowjetunion* (Berlin: epubli, 2015), 220–21). In 1958, Artek came under the control of the Komsomol (RGASPI, f. M – 1, op. 3, d. 981, l. 10).
3 GARF, f. R3341, op. 6, d. 218, ll. 89–90, 110, 113, 115; GARF, f. R3341, op. 6, d. 40, l. 1; Zinovii Solov'ev, "Krym – pioneram!," in *Lager' v Arteke*, ed. CK Obshchestvo Krasnogo Kresta RSFSR (Moscow: Izd. CK Obshchestvo Krasnogo Kresta RSFSR, 1926), 3–8.
4 "Luchshie edut v 'Artek'," *Pionerskaia Pravda*, 4 June 1935, no. 72 (1542): 3.
5 Eichler, *Pionierlager*, 226–28.
6 For proletarian internationalism and Soviet cultural diplomacy in the 1920s and 1930s, see Catriona Kelly, "Defending Children's Rights, 'In Defense of Peace': Children and Soviet Cultural Diplomacy," *Kritika: Explorations in Russian & Eurasian History* 9, no. 4 (2008): 711–46; Michael David-Fox, *Showcasing the Great Experiment: Cultural Diplomacy and Western Visitors to the Soviet Union, 1921–1941* (Oxford: Oxford University Press, 2012).

7 Tobias Rupprecht, *Soviet Internationalism after Stalin: Interaction and Exchange between the USSR and Latin America during the Cold War* (Cambridge: Cambridge University Press, 2015), 1–10. For Soviet internationalism after Stalin's death, see also Eleonory Gilburd, "The Revival of Soviet Internationalism in the Mid to Late 1950s," in *The Thaw: Soviet Society and Culture during the 1950s and 1960s*, ed. Denis Kozlov and Eleonory Gilburd (Toronto: University of Toronto Press, 2013), 362–401.

8 Visiting a capitalist country was much more difficult than visiting a socialist one; however, in both cases, Soviet citizens faced numerous restrictions and had to overcome a great many bureaucratic obstacles. Anne E. Gorsuch, *All This Is Your World: Soviet Tourism at Home and Abroad After Stalin* (Oxford: Oxford University Press, 2011), 81–85, 111–13; Igor' Orlov and Aleksei Popov, *Skvoz' "zheleznyi zanaves": Russo turisto: Sovetskii vyezdnoi turizm, 1955–1991* (Moscow: Izd. dom Vysshei shkoly ėkonomiki, 2016), 44–60.

9 In general, these agreements established exchange programmes for science, technology, the arts, and sport as well as for tourism. Nigel Gould-Davies, "The Logic of Soviet Cultural Diplomacy," *Diplomatic History* 27, no. 2 (2003): 207–8.

10 RGASPI, f. M – 8, op. 1, d. 590, ll. 1–5.

11 The metaphor of the Trojan horse is also used in two articles, Susan E. Reid, "Who Will Beat Whom? Soviet Popular Reception of the American National Exhibition in Moscow, 1959," *Kritika: Explorations in Russian & Eurasian History* 9, no. 4 (2008): 862; and Pia Koivunen, "Overcoming Cold War Boundaries at the World Youth Festivals," in *Reassessing Cold War Europe*, ed. Sari Autio-Sarasmo and Katalin Miklóssy (London: Routledge, 2011), 188.

12 Erving Goffman, *Wir alle spielen Theater. Die Selbstdarstellung im Alltag*, trans. Peter Weber-Schäfer (Munich: Piper, 2011), 104, 117.

13 Dean MacCannell, "Staged Authenticity: Arrangements of Social Space in Tourist Settings," *American Journal of Sociology* 79, no. 3 (1973): 589–603.

14 For the interviews, see the References. Interviews were conducted in Russian and German and are quoted here in my translation. All names have been anonymized.

15 Alessandro Portelli, "What Makes Oral History Different," in *The Oral History Reader*, ed. Robert Perks and Alistair Thomson (Abingdon: Routledge, 2016), 54 argues that an interviewee's memory "is not a passive depository of facts, but an active process of creation of meanings." Therefore, where appropriate I offer a brief account of each respondent's biographical and contemporary social background.

16 Manfred Hildermeier, *Geschichte der Sowjetunion 1917–1991: Entstehung und Niedergang des ersten sozialistischen Staates* (Munich: C. H. Beck, 1998), 176–82.

17 From 1924–1936, this passage was part of the Soviet Pioneer oath (Kelly, "Defending," 723).

18 Ibid., 722–23.

19 Hildermeier, *Geschichte*, 176–82.

20 Kelly, "Defending," 724.

21 Rupprecht, *Soviet Internationalism*, 9, original emphasis.

22 Kelly, "Defending," 726.

23 S. Kaltakhchian, "Internacionalizm," in *Bol'shaia Sovetskaia Enciklopediia*, ed. A. Prokhorov (Moscow: Izd. Sovetskaia Enciklopediia, 1972), 330–31.

24 Ted Hopf, *Reconstructing the Cold War: The Early Years, 1945–1958* (Oxford: Oxford University Press, 2012), 199.

25 Pia Koivunen, "Friends, 'Potential Friends' and Enemies: Reimagining Soviet Relations to the First, Second, and Third Worlds at the Moscow 1957 Youth Festival," in *Socialist Internationalism in the Cold War: Exploring the Second World*, ed. Patryk Babiracki and Austin Jersild (Cham: Springer International, 2016), 221.

26 RGASPI, f. M – 2, op. 1, d. 17, ll. 2–3; RGASPI, f. M – 8, op. 1, d. 590, ll. 1–3; RGASPI, f. M – 1, op. 68, d. 1127s, ll. 65–69.

27 See, for instance, a Komsomol letter to the GDR's Pioneer Organization "Ernst Thälmann" from April 1969 (SAPMO BArch, DY 25/2224, no pagination).
28 Ibid.; SAPMO BArch, DY 24/8695, no pagination.
29 Interviews with Gisela, Peter, and Monika.
30 Pioneers from the GDR used this chance to demonstrate their country's strong economic development. Interestingly, they even provided exhibition visitors with data such as the number of washing machines, TVs, and refrigerators per capita (SAPMO BArch, DY 25/2162, no pagination).
31 SAPMO BArch, DY 25/2224, no pagination; Interview with Gisela.
32 Aleksei Yurchak, *Everything Was Forever, Until It Was No More: The Last Soviet Generation* (Princeton: Princeton University Press, 2005), 58–59.
33 Ibid., 59.
34 SAPMO BArch, DY 25/963, no pagination.
35 Ibid.
36 RGASPI, f. M – 2, op. 1, d. 17, ll. 17–30, 32–51.
37 Ibid., ll. 24–25.
38 Ibid., ll. 23–24.
39 Catriona Kelly, *Children's World: Growing Up in Russia, 1890–1991* (New Haven: Yale University Press, 2007), 150.
40 Ibid.
41 For the relationship between the Soviet collective and the individual, see Oleg Kharkhordin, *The Collective and the Individual in Russia: A Study of Practices* (Berkeley & Los Angeles: University of California Press, 1999).
42 RGASPI, f. M – 2, op. 1, d. 17, l. 22.
43 Kelly, *Children's World*, 120–23, 530; Kelly, "Defending," 724–26.
44 Katalin Miklóssy and Melanie Ilic, "Introduction. Competition in State Socialism," in *Competition in Socialist Society*, ed. Katalin Miklóssy and Melanie Ilic (London: Routledge, 2014), 1.
45 Ibid., 2.
46 Ibid.
47 Ibid.
48 SAPMO BArch, DY 25/963 (no pagination).
49 SAPMO BArch, DY 25/2224 (no pagination).
50 Interview with Peter, 4 April 2017. Peter, from Gelsenkirchen (FRG), accompanied a delegation of the Falcons to Artek in the summer of 1977. Today he works as a child and youth psychotherapist in Gelsenkirchen.
51 According to Pia Koivunen, "The World Youth Festival as an Arena of the 'Cultural Olympics': Meanings of Competition in Soviet Culture in the 1940s and 1950s," in *Competition in Socialist Society*, ed. Katalin Miklóssy and Melanie Ilic (London: Routledge, 2014), 132–33, 135 the same held true for the World Youth Festivals, where one could observe a "significant gap between the contestants from socialist and capitalist countries" too.
52 Interview with Peter.
53 Diane P. Koenker, *Club Red: Vacation Travel and the Soviet Dream* (Ithaca: Cornell University Press, 2013), 4, 7, 15.
54 Interview with Monika, 5 April 2017. Monika, from Gelsenkirchen (FRG), was a member of the Falcons and visited Artek as a teenager in the summer of 1978. She became a pedagogue and works in Gelsenkirchen.
55 RGASPI, f. M – 2, op. 1, d. 17, ll. 27, 42.
56 Nataliia Lebina, *Sovetskaia povsednevnost': normy i anomalii: Ot voennogo kommunizma k bol'shomu stiliu* (Moscow: Novoe Literaturnoe Obozrenie, 2015), 143–44.
57 Yurchak, *Everything*, 168–69.
58 Juliane Fürst, *Stalin's Last Generation: Soviet Post-War Youth and the Emergence of Mature Socialism* (Oxford: Oxford University Press, 2011), 72–73, 193, 200–1, 213, 217, 224.

59 Yurchak, *Everything*, 163.
60 Interview with Gisela.
61 SAPMO-BArch, DY 25/2224 (no pagination).
62 Anna Kozlova, "'Fairy Tale for Pioneers': Deconstruction of Official Ideologies in Memories About Artek 1960s – 1980s," *European Education* 48, no. 3 (2016): 179.
63 Interview with Inna, 8 April 2015. Inna, originally from the Middle Volga Region, visited Artek as a Pioneer in May/June 1977. Today she is a university professor in Russia.
64 Interview with Boris, 28 July 2016. Boris, originally from Central Russia, visited Artek as a Pioneer in the autumn of 1986 or 1987. Today he is a university professor in Russia.
65 Interview with Monika.
66 Interview with Grigorii, 27 May 2015. Grigorii, originally from the Russian Far East, visited Artek as a Pioneer in the spring of 1974. He studied English language and literature and worked as a translator, Pioneer leader, and camp director between 1978 and 1988 for Artek, and then for the Central Council of the Pioneer Organization in Moscow. Since 1990, he has worked for an international youth exchange organization. Artek was a complex of five camps: Sea, Coast, Mountain, Azure, and Cypress. Two of them, Coast and Mountain, were further subdivided into four and three camps respectively. Camp Diamond was part of Camp Mountain (Eichler, *Pionierlager*, 274).
67 Interview with Peter; Interview with Monika.
68 Uwe Engelbrecht, "Was in Artek 'prima' und was 'doof' ist," *Stuttgarter Zeitung*, 6 August 1977. I would like to thank the interviewee Peter who sent me the article.
69 Arne Winkelmann, "Das Pionierlager Artek: Realität und Utopie in der sowjetischen Architektur der sechziger Jahre" (PhD diss., Bauhaus-Universität Weimar, 2003), 77.
70 Interview with Peter.
71 Interview with Monika.
72 Interview with Svetlana, 4 June 2015. Svetlana, born in Central Russia, visited Artek as a Pioneer in the summer of 1982. She studied pedagogy and worked as a Pioneer leader at Artek between 1987 and 1989. Today she is an independent entrepreneur in Russia.
73 Interview with Grigorii.
74 Kozlova, "'Fairy Tale for Pioneers'," 170, 183.
75 For top Komsomol officials widely (mis)using their positions to pursue their self-interests, see Steven L. Solnick, *Stealing the State: Control and Collapse in Soviet Institutions* (Cambridge, MA: Harvard University Press, 1998), 60–124; and Yurchak, *Everything*, 158–237.
76 Interview with Grigorii.
77 Gorsuch, *All This*, 49–78, at 50 uses the term to define the role Estonia played in Soviet tourism. Due to its Western cultural and architectural heritage as well as its European lifestyle, the Baltic republic gave Soviet tourists a feeling of being abroad.
78 Interview with Oksana, 27 May 2015. Oksana, from the Soviet Baltic region, visited Artek as a Pioneer in the summer of 1961. Today she is a university professor in Russia.
79 Ibid.
80 Interview with Grigorii.
81 Anne E. Gorsuch, "Time Travellers: Soviet Tourists to Eastern Europe," in *Turizm: The Russian and East European Tourist under Capitalism and Socialism*, ed. Anne E. Gorsuch and Diane P. Koenker (Ithaca: Cornell University Press, 2006), 216 attributes the expression to an incident during an "evening of friendship" arranged for a group of Soviet tourists in Hungary. One of them preferred to sit with the Hungarian musicians instead of staying with his own group – a decision he was harshly criticized for afterwards.
82 Interview with Grigorii.
83 Interview with Peter; interview with Monika.

References

David-Fox, Michael. *Showcasing the Great Experiment: Cultural Diplomacy and Western Visitors to the Soviet Union, 1921–1941*. Oxford: Oxford University Press, 2012.

Eichler, Evelin. *Pionierlager in der Sowjetunion*. Berlin: epubli, 2015.

Engelbrecht, Uwe. "Was in Artek 'prima' und was 'doof' ist." *Stuttgarter Zeitung*, 6 August 1977.

Fürst, Juliane. *Stalin's Last Generation: Soviet Post-War Youth and the Emergence of Mature Socialism*. Oxford: Oxford University Press, 2011.

Gilburd, Eleonory. "The Revival of Soviet Internationalism in the Mid to Late 1950s." In *The Thaw: Soviet Society and Culture during the 1950s and 1960s*, edited by Denis Kozlov and Eleonory Gilburd, 362–401. Toronto: University of Toronto Press, 2013.

Goffman, Erving. *Wir alle spielen Theater. Die Selbstdarstellung im Alltag*, trans. Peter Weber-Schäfer. Munich: Piper, 2011.

Gorsuch, Anne E. "Time Travellers: Soviet Tourists to Eastern Europe." In *Turizm: The Russian and East European Tourist under Capitalism and Socialism*, edited by Anne E. Gorsuch and Diane P. Koenker, 205–26. Ithaca: Cornell University Press, 2006.

———. *All This Is Your World: Soviet Tourism at Home and Abroad After Stalin*. Oxford: Oxford University Press, 2011.

Gould-Davies, Nigel. "The Logic of Soviet Cultural Diplomacy." *Diplomatic History* 27, no. 2 (2003): 193–214.

Hildermeier, Manfred. *Geschichte der Sowjetunion 1917–1991: Entstehung und Niedergang des ersten sozialistischen Staates*. Munich: C. H. Beck, 1998.

Hopf, Ted. *Reconstructing the Cold War: The Early Years, 1945–1958*. Oxford: Oxford University Press, 2012

Kaltakhchian, S. "Internacionalizm." In *Bol'shaia Sovetskaia Enciklopediia*, edited by A. Prokhorov, 330–31. Moscow: Izd. Sovetskaia Enciklopediia, 1972.

Kelly, Catriona. "Defending Children's Rights, 'In Defense of Peace': Children and Soviet Cultural Diplomacy." *Kritika: Explorations in Russian and Eurasian History* 9, no. 4 (2008): 711–46.

———. *Children's World: Growing Up in Russia, 1890–1991*. New Haven: Yale University Press, 2007.

Kharkhordin, Oleg. *The Collective and the Individual in Russia: A Study of Practices*. Berkeley & Los Angeles: University of California Press, 1999.

Koenker, Diane P. *Club Red: Vacation Travel and the Soviet Dream*. Ithaca: Cornell University Press, 2013.

Koivunen, Pia. "Overcoming Cold War Boundaries at the World Youth Festivals." In *Reassessing Cold War Europe*, edited by Sari Autio-Sarasmo and Katalin Miklóssy, 175–92. London: Routledge, 2011.

———. "The World Youth Festival as an Arena of the 'Cultural Olympics': Meanings of Competition in Soviet Culture in the 1940s and 1950s." In *Competition in Socialist Society*, edited by Katalin Miklóssy and Melanie Ilič, 125–41. London: Routledge, 2014.

———. "Friends, 'Potential Friends,' and Enemies: Reimagining Soviet Relations to the First, Second, and Third Worlds at the Moscow 1957 Youth Festival." In *Socialist Internationalism in the Cold War: Exploring the Second World*, edited by Patryk Babiracki and Austin Jersild, 219–47. Cham: Springer International, 2016.

Kozlova, Anna. "'Fairy Tale for Pioneers': Deconstruction of Official Ideologies in Memories About Artek 1960s – 1980s." *European Education* 48, no. 3 (2016): 170–86.

Lebina, Nataliia. *"Sovetskaia povsednevnost": Normy i anomalii. Ot voennogo kommunizma k bol'shomu stiliu*. Moscow: Novoe Literaturnoe Obozrenie, 2015.

MacCannell, Dean. "Staged Authenticity: Arrangements of Social Space in Tourist Settings." *American Journal of Sociology* 79, no. 3 (1973): 589–603.

Miklóssy, Katalin, and Melanie Ilič. "Introduction: Competition in State Socialism." In *Competition in Socialist Society*, edited by Katalin Miklóssy and Melanie Ilič, 1–9. London: Routledge, 2014.

Orlov, Igor', and Popov, Aleksei. *Skvoz' "zheleznyi zanaves": Russo turisto: Sovetskii vyezdnoi turizm, 1955–1991*. Moscow: Izd. dom Vysshei shkoly ekonomiki, 2016.

Portelli, Alessandro. "What Makes Oral History Different." In *The Oral History Reader*, edited by Robert Perks and Alistair Thomson, 48–57. Abingdon: Routledge, 2016.

Reid, Susan E. "Who Will Beat Whom? Soviet Popular Reception of the American National Exhibition in Moscow, 1959." *Kritika: Explorations in Russian and Eurasian History* 9, no. 4 (2008): 855–904.

Rupprecht, Tobias. *Soviet Internationalism after Stalin: Interaction and Exchange between the USSR and Latin America during the Cold War*. Cambridge: Cambridge University Press, 2015.

Rybinskii, Evgenii. *Pionerskaia respublika Artek: Fotoal'bom*. Moscow: Planeta, 1976.

Solnick, Steven L. *Stealing the State. Control and Collapse in Soviet Institutions*. Cambridge, MA: Harvard University Press, 1998.

Solov'ev, Zinovii. "Krym – pioneram!" In *Lager' v Arteke*, edited by C.K. Obshchestvo, 3–8. Moscow: Izd. CK Obshchestvo Krasnogo Kresta RSFSR, 1926.

Winkelmann, Arne. "Das Pionierlager Artek: Realität und Utopie in der sowjetischen Architektur der sechziger Jahre." PhD diss., Bauhaus-Universität Weimar, 2003.

Yurchak, Alexei. *Everything Was Forever, Until It Was No More: The Last Soviet Generation*. Princeton: Princeton University Press, 2005.

5 Foreign tourists, domestic encounters

Human rights travel to Soviet Jewish homes

Shaul Kelner

The first impression Marge Gordon had of Leningrad was of a "beautiful city with its many bridges and canals formed by Czarist buildings and palaces."[1] Arriving there with her husband Bob on a Monday afternoon in July 1975, the Boston-area artist found the summer weather to be "cool and changeable; one minute it would be bright and sunny; the next minute we would be caught in a sudden downpour." She was glad to have brought her compact umbrella. Peter the Great's imperial city made a better impression than Moscow, the first stop on the couple's two-week visit to the USSR. Introducing the Soviet capital at the start of her five-page travelogue, Gordon, with a painter's preference for colour and beauty, and a palette full of American Cold War tropes, wrote,

> Moscow struck me as being a cold, gray, impersonal and unaesthetic city. The people on the streets seemed colorless, and drab, their expressions blank and even hostile. No one was helpful to us, seemed friendly or even curious towards us as foreigners. Everywhere there were cues [*sic*] – lines of people waiting in stores, for taxis, and of course there was a tremendous long cue to go into Lenin's Mausoleum.[2]

This is about all that Gordon wrote about the cityscapes. Although she and her husband toured the major attractions, sightseeing was incidental to their visit. "We had decided to spend the first day in each city with our tour group; so Tuesday July 8 was spent on a bus tour. . . . By Tuesday evening, we were anxious to begin the real purpose of our trip – to make contact with Jewish activists."[3] Over the next ten days in Moscow, Leningrad, and Kiev, the couple made 16 separate visits to the apartments of nine different families who were involved in the struggle for Soviet Jewish emigration rights. Some they visited as many as three times.[4] In most instances, the visits brought the Gordons into contact not only with those who lived in the apartments, but also with other Soviet Jewish activists who came round to talk with the Americans.

Gordon was one of thousands of Westerners who visited the Soviet Union in the 1970s and 1980s as part of a human rights campaign to aid persecuted Jews who had been denied the right to emigrate (known as *otkazniki* in Russian and refuseniks in English). Her travelogue, solicited by and filed with an American advocacy group for Soviet Jewish emigration rights, is now held in its archives along with thousands of similar journals and trip reports.[5] In a manner that is representative of this unique set of Cold War-era testimonies, Gordon's travelogue devotes only minimal space to the postcard-like scenes of Red Square and the Hermitage, focusing instead on the intimate spaces of domestic life, where individuals shared their personal stories. Other than her brief comments about Leningrad's beauty and her short aside on the weather, Gordon said no more about the city in general. Instead, her narrative shifted to describe spaces that few tourists ever saw:

> Tuesday evening we went to see Zhanna K. who [the refusenik] Kim Fridman in Kiev recommended we see. Zhanna lives in a tiny apartment and shares the kitchen and bathroom with a married couple who become violent when drunk which is most of the time. Zhanna had told us that her last American visitor had a chair thrown at him from three stories up by her drunken neighbors. We were more fortunate as there was no sign of them while we were there.[6]

The "tourist gaze" has been understood as an act of semiotic engagement, a reading of places as symbols of themselves in an effort to make meaning of them.[7] During the Cold War era, Western travellers used visits to the Soviet Union as a means of directing the tourist gaze at the Cold War itself, using the country as a theatre of signifiers of the geopolitical conflict that divided the globe. But whereas the majority of Western sightseers in the USSR engaged in this Cold War tourism only in their consumption of Soviet public space, Marge Gordon and other human rights tourists constructed their understanding by turning their gaze on Soviet domestic spaces as well. Thus, while Gordon's comment about the huge queues for Lenin's Mausoleum reproduced an image of Russia so common among American travellers that it even found its way into children's books,[8] she was also able to turn her tourist gaze away from the glass casket that displayed Lenin's remains and make a tourist attraction out of a different glass case, hidden from public view:

> As we all sat across from Zhanna in her tiny, rectangular room, I gazed at the similarly proportioned rectangular fish tank above her minuscule table. The large fish swimming around and around reminded me of those Refuseniks similarly entrapped.[9]

Detente, tourism, and Soviet domestic space

In 1974, one year before the Gordons' visit, Fodor's published its first travel guide to the Soviet Union. The superpowers were pursuing detente, and policymakers and industry representatives alike were touting the notion that tourism could forge human relationships to undergird the government rapprochement.[10] Fodor's editors took up the "peace through tourism" refrain in the introduction to their new volume:

> This is, after all, the season of détente, an ideal which we heartily endorse. We believe it essential for those of us engaged in encouraging tourism to

support all sincere attempts at People-to-People contact, no matter where or with whom, and specifically without regard to political differences between nations. We rejoice at the prospect of the lessening of tensions, as we feel this is conducive to more travel and that travel means, we think, more understanding.[11]

Similar claims were enshrined in Helsinki the following year, as the signatories to the Final Act of the Conference on Security and Cooperation in Europe (CSCE) committed themselves to "encourage increased tourism" based on the recognition "of the contribution made by international tourism to the development of mutual understanding among peoples." It was a point reinforced for Soviet audiences in the pages of *Literaturnaia Gazeta* in 1976 by the head of Intourist, Sergei Nikitin, who praised tourism for its role in fostering a "material reduction of tensions" between East and West.[12]

Technology and commerce conspired to translate detente's political imperative into an actual flow of sightseers across Cold War borders. The British airline BOAC had brought commercial transatlantic travel into the jet age in 1958, and ten years later, in July 1968, Pan Am and Aeroflot signed a deal to inaugurate direct flights between New York and Moscow.[13] The arrangement helped increase the number of American visitors to the USSR by over 700 per cent, from 14,000 in 1964 to 76,000 in 1970, and on to 114,000 in 1975, and the number of Soviet visitors to the US from 12,000 in 1970 to 28,000 in 1975.[14]

Although the rhetoric associated with this surge in cross-bloc travel celebrated tourism as a way of promoting mutual understanding and solidifying detente, the reality was more complex. Even after reopening the USSR to Western tourism in 1955, Soviet leaders never shed their concerns about the corrupting influence of contact with Westerners and of the potential for the US and its allies to use tourism as a channel for espionage. Soviet tourism authorities maintained tight controls on travellers going abroad and on foreign travellers entering the USSR in order to minimise the possibility of unsupervised contact between Soviet citizens and Westerners.[15] Their concerns were not unfounded. In the detente years, which coincided with the so-called Brezhnevian Stagnation, exposure to the West did help undermine Soviet citizens' confidence in their own system.[16] Moreover, Western governments were as inclined as the Kremlin to enlist tourists and participants in cultural exchanges as agents of the state, using them to engage in cultural diplomacy and to quietly gather information from and about their Cold War rivals.[17]

As for the notion of tourism as the handmaiden of detente, perhaps nowhere was the gap between rhetoric and reality more evident than in the fact that human rights groups seized on the expansion of tourism to move Western activists into the USSR. Once there, travellers did indeed strike up people-to-people relationships with Soviet citizens, but not as detente's paeans to tourism had envisioned. Instead, human rights tourists built ties with dissidents, would-be émigrés, and victims of government abuse determined to make their voices heard in the West. From the mid-1970s onwards, the transnational alliances that tourism helped forge between human rights activists on both sides of the Iron Curtain ratcheted up the pressure for liberalising reforms that, once underway, culminated in the fall of communism.[18] It was one of the ironies of detente that tourism, imagined as an interpersonal practice that would help stabilise the superpower relationship and create a durable, manageable Cold War equilibrium, instead had a destabilising effect that helped hasten the collapse of the USSR, bringing the Cold War to an end.

This is the end of the story as far as scholars of international relations are concerned. For sociologists and social historians trying to understand the diversity of Cold War culture, by contrast, human rights tourism is notable less for any contribution to the fall of communism it may have had, than for the fact that for two decades it provided thousands of ordinary Westerners the rarest of vantage points into Soviet life: the view from inside Soviet homes. Even in an era when Westerners' understandings of the Soviet Union could increasingly be informed by first-person experiences made possible by tourism, most tourists' impressions of the USSR came from viewing its public space. Politically driven restrictions on contact between tourists and locals were partly responsible for this, but much of it was also due to the economic efficiencies of package tourism. Circumventing Intourist, however, Western organisations working for Soviet Jewish emigration rights collaborated with Jewish activists in the USSR to create an alternative tourist track that regularly opened Soviet apartments to Western visitors. Prior to the tourists' departure, the organisations provided them with Soviet Jews' addresses and phone numbers and instructed them on how to break away from the package tours, use Soviet payphones to initiate contact, and navigate the Soviet metro and taxi systems. Contrary to the hopes of those who saw tourism as a means of fostering goodwill, the encounters in Soviet apartments tended to reinforce rather than mitigate the negative Cold War imagery of the Soviet Union.

Activism and tourism

When the Gordons visited the Soviet Union, they did so with the help of Action for Soviet Jewry, a Boston-area human rights organisation that campaigned for persecuted Soviet Jews (Bob Gordon was a co-founder of the group). Such organisations were found throughout the West, wherever sizeable Jewish communities existed. The groups constituted one-third of a transnational campaign, the others being the informal networks of Jewish activists in the USSR and a semi-clandestine agency of the Israeli government, known as Nativ (Hebrew for "Pathway" or "Route").[19] The first Soviet Jewry advocacy organisation in the US, the Cleveland Committee on Soviet Anti-Semitism, had been founded in 1963 by members of a synagogue book club in Ohio. By the time of the Gordons' trip in 1975 (just weeks before the Helsinki Accords were signed), the network of organisations had spread around the world, and had succeeded in raising public awareness and mobilising Western government opposition to anti-Jewish policies in the USSR. And the list was long. There were bans on the baking of unleavened bread (*matzah*) for Passover; government publications vilified the Jewish religion as "racist," "reactionary," "anti-Soviet," and "hostile to the fundamentals of socialist morals;"[20] all but a token few Jewish cultural institutions – synagogues, theatres, newspapers, and schools – had been closed;[21] there were Jewish quotas in universities; there was workplace discrimination; and there were reprisals against those who sought to escape their situation by applying for permission to emigrate (loss of work, police harassment, arrest, and imprisonment).[22] As a result of an energetic effort by the transnational campaign, what activists referred to as the "plight" of Soviet Jewry made newspaper headlines, was testified to in Congressional hearings, negotiated over in US–Soviet diplomatic contacts, and shouted about in public protests, the largest of which – a December 1987 rally in Washington, DC – numbered over a quarter of a million.[23]

The Western campaign for Soviet Jews had two strands: supporting Jewish culture inside the USSR and helping Jews emigrate. It was an effort that brought NGOs and Western governments together in partnership.[24] The Israeli government played a crucial role in launching the campaign, establishing Nativ in 1952 to provide direct support for Soviet Jews and to foster grassroots activism in other countries throughout the West.[25]

Nativ was the first of the organisations to recognise tourism's potential as a resource, enlisting it to open a line of communication with Jews on the far side of the Iron Curtain. In 1966, it began supporting efforts to send Israeli citizens as "tourist-emissaries." After the Soviets severed diplomatic relations with Israel as a result of the Arab–Israeli Six Day War in June 1967, Nativ turned to recruiting Jewish tourists from Europe and North America instead, sending its first groups from London and Stockholm in late 1967, from Paris in spring 1968, and from New York that autumn.[26] Within two to three years, NGOs such as the New York-based National Conference on Soviet Jewry (NCSJ), the Union of Councils for Soviet Jews (headquartered at the time in Cleveland), the Women's Campaign for Soviet Jewry in London, the Comité des Quinze in Paris, the Aktionskommittén för Sovjets judar in Stockholm, and others began following suit. The Union of Councils began contemplating systematic efforts to mobilize tourism in 1970, with initial attempts at implementation beginning in late 1971. The NCSJ announced its own tourism effort in January 1972.[27]

The NGOs worked with three main types of travellers: the likes of the Gordons, who were travelling at the behest of the organisations for the primary purpose of meeting Soviet Jews; members of government and official delegations who met Soviet Jews as part of their missions; and general travellers (holidaymakers, conference-goers, students, etc.) whose visit had nothing to do with Soviet Jewry, but who were interested, or at least willing, to meet local Jews while in the USSR. Some of the latter sought out Soviet Jewry NGOs for assistance, while others were recruited when activists learned of their travel plans.

Although Russian-speaking tourists were prized recruits, they were few in number. As a result, conversations tended to be held in a combination of Western and Jewish languages, with running translations into and out of Russian provided by multilingual refuseniks. Yiddish, which had been the vernacular of Eastern European Jewry until the war, was the shared language for older tourists and refuseniks, who had learned some of the language from their parents and grandparents.[28] The younger generations relied more on Modern Israeli Hebrew, a language that, if spoken at all, would typically have been acquired by hosts and guests alike as a second language.[29] Most American travellers spoke in English, as the refuseniks' English tended to be better than the visitors' Russian. English-speakers with a knowledge of French and German found these languages useful as well.

To support the travellers and facilitate their contact with Soviet Jews, NGOs prepared guidebooks with instructions on how to find synagogues, use payphones, decode the Cyrillic alphabet, tip taxi drivers, navigate the metro, smuggle things through customs, and slip away from tour groups without arousing suspicion.[30] Activists briefed and debriefed travellers, providing them with refuseniks' names, phone numbers, addresses, and biographies, along with directions to their apartments from the nearest metro stations. They also provided tailored lists of items needed by the people they were to visit – tourists helped refuseniks by bringing in books and medicine, office supplies to support local activism, and goods to sell on the black market (since refuseniks were often struggling financially, having been fired from their jobs).[31] Tourists also helped by bringing information out, often in the form of trip reports ranging from a few paragraphs to a hundred pages in length. The information provided by returning tourists enabled the NGOs to compile databases with details of each refusenik's legal status, health, material needs, and more. There are thousands of these trip reports in the archives, the bulk of which – including Marge Gordon's – are now held by the American Jewish Historical Society's Archive of the American Soviet Jewry Movement.[32]

Despite the platitudes about tourism reducing Cold War stereotypes and fostering goodwill, the mobilisation of tourism by the Western campaigns for Soviet Jews tended to accomplish the opposite. The standard practice in this type of tourism involved vanishing from the organised Intourist groups in order to meet refuseniks in their homes. There, conversations centred on the hosts' desire to emigrate (usually to Israel) and their negative experiences with the government, which both motivated and resulted from their efforts to leave.

Customs agents, Intourist guides, the police, and the KGB were aware of what tourists were doing, and tourists usually suspected and sometimes knew that they were being watched. Most travellers passed through customs on arrival and departure without encountering problems, but this ran so counter to their expectations that travellers were surprised at how easy it was. "For unexplainable reasons," one traveller wrote, "we walked through customs without being stopped or examined. It proved that 'in Russia, the rule is that there is no rule.' You never know when you are [going to be] taken apart at customs or simply ignored."[33] In many instances, however, tourists did, in fact, have difficult encounters with Soviet officials. The most common report was of being singled out for invasive searches and harsh questioning at customs. Some also reported being followed by plainclothes police, or being harassed in their hotel rooms (receiving midnight calls with no voice at the other end of the line, or returning to their rooms to find that their luggage had been opened and searched). In a handful of instances – not more than

5 per cent of the trips reported on – people were arrested and interrogated. A few of these individuals were expelled from the country, and some of the expulsions were followed by denunciations by name in the Soviet press.[34]

Unrepresentative populations

Scholars who have looked at Westerners' visits to communist-bloc homes have concentrated on who had the privilege of taking part, for neither hosts nor guests were representative of their countries' broader populations. State-sponsored city-twinning programmes between Soviet satellite countries and Western states created some opportunities for municipal leaders to be hosted by their Eastern European counterparts.[35] Exchange programmes with the Soviet Komsomol youth movement gained some young people from the West a glimpse of life in Soviet apartments or dormitories.[36] In the 1970s, Intourist began offering itineraries that included "structured 'evenings of friendship'" for "freely improvised" conversations on particular themes.[37] As for privately initiated, unstructured domestic encounters, these were class-biased: Western participants tended to come from an elite group of journalists, diplomats, and academics in residence in the USSR; hosts, from an equivalent social stratum of educated Soviet professionals, primarily in Moscow and Leningrad, including government officials and human rights activists (though probably not at the same gatherings).[38] Westerners in the Soviet Union on short-term tourist visas generally did not find homes opened to them in this manner. Human rights tourists were a notable exception.

Soviet Jewry movement tourism expanded the range of Westerners who had a glimpse of Soviet home life, even as the profile of its travellers remained distinctive: business people, teachers, non-profit workers, clergy, doctors, and lawyers, most of them Jewish.[39] Their Soviet hosts, being Jewish activists pressing to emigrate to Israel, were in their own way unrepresentative of the broader population (and even of the broader Soviet Jewish population, most of whom were not activists, and who, when given the choice, tended to opt for the US over Israel).[40] While most Soviet Jewish hosts were from Moscow and Leningrad, many were not. Of the 3,500 or so tourists whose trips from the 1970s and 1980s are accounted for in Soviet Jewry NGO files, over 90 per cent visited refuseniks in Moscow, about 85 per cent in Leningrad. A quarter went to Kiev. Overall, slightly under half (approximately 45 per cent) travelled beyond these three cities to visit refuseniks in cities across the Ukraine, the Baltics, Georgia, and the Caucasus. Tbilisi, Odessa, Riga, and Vilnius each received about 8 per cent of the visitors (between 275–300 people each), Minsk 7 per cent, Tashkent 5 per cent, and Samarkand, Yerevan, and Tallinn about 4 per cent each (around 130 people).[41]

By providing contacts in the USSR and guidance on arranging meetings with refuseniks, the Soviet Jewry movement created one of the few modes of tourism that regularly opened Soviet domestic spaces to Americans and Western Europeans visiting the country on short-term tourist visas. Moreover, by encouraging them to write up their encounters, the movement extended the immediate meaning-making of the tourist gaze into a more sustained process of second-order reflection via narration of the travel experience. Thousands of normal tourists took up their pens and became, for a moment, Cold War era travel writers, their representations of the Soviet Union informed by experiences inside Soviet homes.[42]

The semiotics of "authenticity" in tourism

In analysing how home visits figured into travellers' representations of the Soviet Union, it is important to remember just how much visits to people's homes occupy a privileged position in the semiotics of international tourism. Tourism is, among other things, a way of learning about foreign places, of making the unfamiliar familiar. Much of the tourism industry is premised on catering to this, representing places and putting meanings on display. But the very fact that an industry has arisen to serve tourists has helped to create the imperative to "escape the tourist bubble" and "get off the beaten track," to see sites untouched by the tourism industry.[43] Of these so-called "authentic" travel experiences, few are more prized than an invitation from locals to join them in their homes. There are sound sociological reasons for this, as Dean MacCannell shows in a classic work that draws on Erving Goffman's dramaturgical notion of front-stage and back-stage regions:

> Having a back region generates the belief that there is something more than meets the eye, *even when there are no secrets actually kept*. . . . Being 'one of them,' [i.e. a local] or at one with 'them,' means, in part, being permitted to share back regions with 'them.' This is a sharing which allows one to see behind the others' mere performances.[44]

In the general context of international tourism's symbolic economy, refuseniks' apartments were well positioned to serve as sites that would enable visitors to feel they were "discovering" an "authentic" Soviet Union. To the extent that it is possible to speak of an "elusive private sphere" in the USSR, apartments were the closest approximation to a back stage where the locals' private lives unfolded.[45] If not entirely beyond the reach of the state, they were at least decidedly beyond the control of the Soviet tourism industry. The travellers' belief that they were "off the beaten path" was informed by the difficulties they encountered in trying to find the apartments. The difficulty of making their way alone to a refusenik's apartment was a common theme in the reports:

> The next morning before breakfast we decided to walk to [refusenik Vladimir] Slepak's flat. There are no accurate and detailed Russian maps available to tourists – probably to discourage people such as we, and on our incomplete map, Slepak's flat appeared to be just on the other side of Red Square. After walking at least one and one half hours without finding his place and feeling cold and discouraged, we returned to the hotel. After breakfast we again set out again and this time finally were successful. We felt such relief when Slepak came to his door and welcomed us into his apartment.
>
> (Marge Gordon, 1975)[46]

I took the Metro for one station, walked the area for an hour, asking directions several times, finally gave up, spent another half hour trying to hail a
cab back to the hotel, walked into my hotel room, disgusted with myself,
disgusted with my American friends who gave me the names and addresses,
disgusted with Soviet Jewry when the phone rang:
  – Hello. This is Ivan. I received your cable.
  – Thank God. I was beginning to think that you were a fictitious character
from a James Bond novel.
  We met two hours later in front of a store, I with an orange kerchief so that
he could identify me.

(Ruth Nordlicht, 1975)[47]

I strongly recommend that the traveler prepare himself before going off alone.
He should spend some time learning the Cyrillic alphabet and a few Russian
phrases. It is otherwise very difficult to wander from the group or make visits
to people's homes. It is also wise to purchase, in advance, maps of the major
cities. American- and European-published maps are available showing street
names and metro stops in both Cyrillic and Roman letters. They will not be
easily available in the USSR.

(Harriet Goldberg, 1979)[48]

And yet representations of Soviet Jewish domestic spaces are contradictory,
because in spite of the aura of authenticity generally ascribed to back regions,
travelogues tended not to use descriptions of apartments to make explicit gen-
eralised assertions about the Soviet Union. This was in marked contrast to
the representations of public space. Travelogues were full of generalising
descriptions – "Most of the streets are teeming with masses of people at all
times"[49] – or guidebook-style overviews like Marge Gordon's – "We were struck
by how different [Kiev] was from Moscow. It was warmer and prettier perhaps
because of the many flowers in bloom, and because of the scenic Dnieper River
with its many bridges. The people seemed more Western to us and more stylish (miniskirts and even platform shoes were not unusual)."[50] Domestic spaces
tended not to be accorded similar treatment. One young attorney's travelogue
from 1979 was a rare exception:

The visits to refuseniks showed another side of things . . . [and] brought to
life some of the things that the Soviet In-Tourist guides discussed. We were
often told, for example, about the acute housing shortage in the Soviet Union
and the national goal of providing nine square meters of space for every citizen. . . . My visits to two Soviet apartments showed me that nine square
meters provides little space and less privacy.

(Harriet Goldberg, 1979)[51]

In fact, one of the most striking things about the representation of domestic space
in the travelogues is how often it was *not* represented at all. In the majority of

travelogues, reports on conversations in refuseniks' homes covered the content of the discussions without offering any description of the physical setting in which the discussions took place. All that readers learn, in most instances, is that so-and-so's apartment was the venue for the meeting.

Behind this tendency to write about the content of a conversation while ignoring its setting was the imperative of genre and audience. The reports were written for organisations whose briefing and debriefing materials clearly communicated a preference for certain types of information over others. The National Conference on Soviet Jewry was one of several organisations that provided travellers with fill-in-the-blank questionnaires: two pages summarising the details of the trip, and two pages each per refusenik visit, to include the city, name of person visited, address, phone number, how contacted, items left, requests for assistance, and items brought out, with space at the end for additional remarks.[52] Descriptions of apartments were not on this list, and one can surmise that travellers understood that the organisations were culling information for their databases.[53] What would be helpful for the movement was to provide information that would be "useful."

What, then, are we to make of the fact that a sizeable minority of travellers did choose to describe the apartments they visited? The answer, I believe, lies in recognising that although Soviet Jewry NGOs only sought specific types of information, they also created an imperative to write, and in so doing set up the conditions that generated more than they had asked for – not merely data but narratives, stories that were not in themselves immediately useful to their information processing purposes. Travellers ended up putting pen to paper not only for the organisations that sent them, but also for themselves. They narrated the stories of their journey as a form of self-expression, making sense of their experience engaging in activism not from afar, but in the presence of refuseniks in the Soviet Union itself.[54]

Descriptions of refuseniks' apartments figured in this sense-making in a variety of ways. They helped travellers grasp the refuseniks' situation as one of professional success abandoned for the hope of a life elsewhere. They also helped travellers discover evidence that could affirm the value of their own presence in the Soviet Union. Perhaps most important, the representation of Soviet Jewish domestic space both as a haven from the Soviet world outside and as penetrated by that world helped travellers reinforce classic Cold War understandings of the Soviet Union as an oppressive society.

Gazing on the other, gazing on the self

Domestic spaces, when they were written about in the reports, tended not to serve as focal points of the narrative. Only occasionally written as symbols of themselves, they were more often presented incidentally, as the setting for interactions and conversations that were the narrator's main focus. It is the writers' casual attention to the issue that makes their accounts of domestic space especially revealing, because one can interpret them as moments of slippage that reveal assumptions and implicit meanings that the writers themselves might not have been aware of in discursive consciousness. As the British humourist, Douglas Adams, put it, "Words used carelessly, as if they did not matter in any serious way, often [allow] otherwise well-guarded truths to seep through."[55]

The collection of travelogues reveals that certain aspects of Soviet domestic space were remarked on by many different travellers. These included, most prominently, hospitality, domestic surveillance, relations with neighbours, and apartment size, upkeep, décor, and furnishings. Not all of this served to represent the Soviet Union per se (although the assemblage of synonyms noting the "small," "pathetically small," "humble," or "tiny" size of the apartments certainly did).[56] Descriptions of décor and furnishings, for instance, mainly served the rhetorical function of characterisation, presenting the apartments' occupants as cultured men and women who were navigating the consequences of their resistance. Bookshelves and their contents were high on the list. Russian books signified high levels of education and professional attainment, the latter now redundant; Hebrew books signified Jewish commitment; English and other foreign-language literature, including Western newspapers, magazines and scientific journals, offered evidence of a lifeline from the West and a determination to remain professionally engaged despite unemployment – signifiers of the refuseniks' success in coping with their situation.

Typically, any such signalling was brief, as in these two July 1975 descriptions of the books in Vladimir and Masha Slepak's Moscow apartment:

> I was also impressed with the English books and publications on his shelves. It seemed that he had a well-stocked up-to-date library.
> (Marge Gordon, 1975)[57]

> They have an extensive English Judaica library, including the most recent copy of 'Commentary Magazine.'
> (Trudy Shecter and Debbie Shecter, 1975)[58]

In rare instances, writers made the implicit explicit, acknowledging that they were treating refuseniks' libraries as signifiers of deeper meaning, as in this stream-of-consciousness, composite portrait written by Rochelle Ginsburg, an elementary school educator who visited a decade later, in 1986:

> Our eyes, moving in every direction at once, work to sort out our surroundings. The things of every household stuffed onto shelves and into corners. A mixture of necessities and mementos. Obvious signs of other visitors – cards, photographs, foreign magazines, collected and displayed. Cups and saucers stacked on shelves behind glass. But, most of all, books. Books lining every shelf and surface and table top available. Books in Russian, Hebrew, English. Textbooks, paperbacks, leatherbound volumes. Technical books, literary works, religious books. How could such small spaces contain so many books? While listening to the voices around me it is hard not to drift into contemplation about the books on the shelves. They tell a kind of story

in themselves. The past. The education and technical training. The level of expertise achieved. These are the students, and in some cases, the authors. The religious books, language primers, prayer books, and commentaries. The entry into a Jewish life that may or may not have developed until recently. The classic literary works and contemporary popular works – contact with a more cosmopolitan world. The current journals and special interest and news magazines – Reassurance that there is a pulse, there is a heartbeat, and they can still feel it.[59]

The "obvious signs of other visitors" mentioned by Ginsburg were also noted by other travellers. In June 1973, a New York ophthalmologist, Dennis Freilich, called attention to the decidedly non-Soviet political slogan decorating the Leningrad apartment of 18-year-old Chanoch (Yevgeny) K.:

> The first thing that could be noticed in the room was a large banner on the wall saying 'Don't Blame Me, I voted for McGovern.' This was given to him by an American student, Seth [A.], who was present there on Passover.[60]

The banner was a visual non sequitur, but Freilich let that pass without comment. Readers in American Soviet Jewry organisations would recognise the slogan and its humour. It was popular with members of the Democratic Party, whose presidential candidate, George McGovern, had just lost in a landslide to the incumbent Republican, Richard M. Nixon, in the elections of November 1972. (American Jews, including those active in the campaign for Soviet Jewry, tended not to look favourably on Nixon.)[61] Six months after Freilich's visit, Jerry Lewis, a former clerk to Britain's All Parliamentary Committee for the Release of Soviet Jewry, offered these observations:

> Most of the flats have maps of Israel on the walls, and postcards from overseas sent by wellwishers displayed on mantelpieces or alongside the maps. They also proudly show the photographs of activists working for them in other countries, and treasured the presents and trinkets sent to them by their overseas wellwishers.[62]

In these descriptions, the conventional relationship between tourist and souvenir was seen to be inverted. While the tourists may have later displayed souvenirs of their travels once they had returned home, during their stay in the Soviet Union they noticed that it was the locals who were commemorating the tourists' visits by displaying mementos left in Russia by other travellers. In this, the tourists not only represented Soviet Jews as successfully remaining hopeful and sustaining social connections in spite of their circumstances, but more important, they were also telling themselves that their visits mattered, and that their support was valued. Travellers saw in Soviet Jewish apartment décor an unspoken affirmation of the importance of their own presence there.

The antithesis of Soviet space

In travellers' accounts of their home visits, refuseniks' apartments were typically presented as sites of warm hospitality. This emphasis took on a broader significance given that the rhetorical function of such descriptions was not to exemplify the Soviet Union writ large as a warm, vibrant, colourful country, but to establish a contrast that reinforced the Cold War imagery of a cold, impersonal, inhospitable place.

The travelogue written by Victor Borden, an obstetrician and gynaecologist from New Jersey who visited the USSR with his wife, Frani, in July 1985, is illustrative. It contains a running theme common to many of the trip reports: the American Cold War stereotype of *homo sovieticus* as a grey, unsmiling automaton. The trope is present even in its opening lines:

> We land at Moscow on July 4th. How ironic that on the day we celebrate our independence, we arrive in the Soviet Union. . . . I notice that none of the arriving passengers from our flight exhibit any joy or happiness at having arrived at Moscow. There seem to be no families waiting to welcome home those arriving passengers. This is our first indication of what we later find to be prevalent in all of Russia. There is no joy.[63]

Several pages on, however, Borden did find families and joy in Russia – not in public, but in the domestic realm, in refuseniks' homes. He described his visit with Frani to the apartment of the activists Alec and Rosa Ioffe, for their son Dmitry's wedding.

> In this small little room, filled with bookcases, approximately thirty people crowd in to witness the wedding. It is oppressively hot and uncomfortable, but the emotions of the moment far exceed the discomfort. We are standing so close that Frani is actually under the tallit [prayer-shawl wedding canopy], and we can easily observe the bride weeping with joy during the ceremony. I cry when I hear Dimitry, in Hebrew, say, "I take thee for my lawful wedded wife." A huge, 'Mazel tov!' erupts from everyone in the room when the glass is broken. . . . Rosa, Alec's wife, has put out large quantities of cake and wine, which we partake of after the ceremony. . . . Rosa serves different salads, deviled eggs, hot peppers, fish, and challah (over which our cover is used).[64] We can see that a lot of preparation was necessary for this meal. Large quantities of wine and Vodka are present. There are at least twenty people sitting along side this table. There is hardly any elbow room for any of us, but no one cares, as we all are thrilled to be able to sit so close to each other and share in these precious few moments together.[65]

The morning after the wedding, the Bordens were back with their tour, out in public spaces, their gaze taking in other people who were crowded together – the anonymous Soviet masses:

After a few hours of sleep, we get up, and have the same breakfast as yesterday. Leaving the hotel, we walk to GUM, a huge department store located near Red Square. . . . There are huge crowds of people throughout the store, many waiting in lines, especially for shoes and cosmetics. We find nothing stylish in any of the departments of the store. As an example, we can find no leather goods, only those of vinyl, yet people are lined up for this terrible merchandise. In order to purchase an item, a customer must stand in three lines, one to look at the item and get its exact price, the second to pay for the item, and the last to pick it up once he shows a receipt. All of this seems to take forever.[66]

The text continues alternating in this way, establishing a binary contrast between Soviet public space and private Jewish domestic space. Unsmiling Russians, an anonymous, undifferentiated crowd, waiting in long queues for low-quality mass-produced goods; joyous Jews, named and individuated, packed together and enjoying home-cooked delicacies in small apartments, havens of warmth and camaraderie in an otherwise cold and unfriendly country. Victor Borden himself made the contrast explicit at various points of his travelogue:

I must now take some time to give some impressions of the Russian people, *not the refuseniks*, we have met. They walk or rather march along as if they are robots. They are expressionless zombies, their demeanor reflecting no pleasure or joy in their existence. They work not to achieve but because they have to.[67]

My G–d, how I miss seeing people smile! The only laughter we've heard *other than at a refusenik's home* has been at the circus. A person can't help but laugh at the circus, but the moment we walked outside everyone again wore grim unhappy expressions.[68]

When Frani mentions that all the people we observe on the street seem emotionless, but the Refuseniks we have met, in spite of their awful predicaments, exhibit a greater variety of emotions, Evgeny responds by saying that once he and the other Refuseniks decided to apply for exit visas they inwardly became free. A great burden was taken from them. It is that inner sense of freedom that makes them better off than the rest.[69]

It is important to recognise that there are other ways the writer might have interpreted the experiences in the refuseniks' apartments. Borden could have concluded that his initial impressions about a joyless Soviet Union were a Cold War stereotype. In fact, there was joy. In fact, there were smiles. He had seen it at first hand. He could have declared that his initial impression that Soviets were joyless resulted from the mistake of looking only at their behaviour in public. He might have suggested that Soviets behaved differently in public and private, being reserved with strangers, effusive with friends. But to draw this conclusion, he would have had to have seen the refuseniks as Soviets just like the strangers on the street.

Why was it so hard to see them in this way? Jews were othered in the USSR, refuseniks especially so, and those trying to help them emigrate took the fact of this othering for granted. Moreover, with no access to the homes of Soviet gentiles, Borden's only experience of them was in public. The result was an interpretation of the experience in Soviet Jewish apartments not as representative of the Soviet Union, but as one half of a set of related dichotomies: Jewish/domestic space/warmth/joy were mapped onto one side; Soviet Russian (gentile)/public space/cold/joyless, the other. In this way, domestic encounters, which under other circumstances might have undermined Western Cold War preconceptions, ended up reinforcing them.

Public encroachments on private space

If the focus on warmth and hospitality showed the Soviet Union in a negative light by portraying refuseniks' apartments as havens from the hostile country beyond, negative representations of this outside world were also communicated in portrayals of these havens as insecure. Visitors commonly represented Soviet Jewish homes as sites where the public sphere penetrated private lives, making domestic space into a site of state surveillance. Sometimes, this was written about by relating refuseniks' reports of police and KGB entry into their homes to conduct searches or make arrests:

> Natasha and her husband, Mikhail, are very active in refusenik activities. He teaches Hebrew, they have met with American Congressmen, and in general, coordinate refusenik battles with officials. Their flat has been searched many times by the KGB with materials occasionally confiscated. They worry constantly about being arrested and imprisoned, but, nevertheless, they continue on with their activities.
>
> (Victor Borden, 1985)[70]

In other instances, travellers portrayed Soviet Jewish domestic space as sites of state surveillance by describing their own sightings of KGB watchers stationed outside or entering apartment buildings.

> Arriving at Professor Lerner's apartment house, we found two KGB agents in the lobby – and a smiling Lerner at his door. . . . We talked freely, if carefully, knowing there were two cars with KGB men keeping surveillance of the building.
>
> (Nat Kameny, 1974)[71]

> [Dorian and Anya H., of Riga] both stated that their phone is tapped and that they are constantly being followed and harassed. I did personally witness a 'plumber' appearing at their apartment at 8:30 p.m. to supposedly fix pipes. They claim they had called for no one, and I can only assume that the transmission of our conversation was not coming through clearly.
>
> (Jules Lippert, 1973)[72]

A third way that travellers represented Soviet apartments as something other than private havens was by describing behaviour by their hosts that seemed to indicate that the refuseniks themselves suspected they were under surveillance. For example, when refuseniks sometimes moved their conversations out of their apartments, travellers were wont to interpret this as a sign that the domestic space had been penetrated by the surveillance state and was unsafe for open discussion. "We went to see Professor Lerner," wrote Ahaviah Scheindlin and Bill Aron in 1981. "We had a good visit. We discussed all meetings, etc. in the park across the street (we couldn't speak in anybody's home in Moscow!)."[73]

At the other extreme were refuseniks who framed their domestic space as thoroughly compromised by surveillance, not by holding their tongues, but by demonstrably making a show of *not* concealing information. Whether this was done with humour or with defiance, the implication was that since it was impossible to keep anything secret from government watchers, homes were no less safe for conversation than anywhere else. Visiting two of the most internationally well-known refuseniks, Vladimir and Masha Slepak, Marge Gordon wrote, "It impressed me that they spoke so freely and openly about their situation even though they assume that their apartment is bugged. In fact, on occasion Slepak would look up at his wall and speak jokingly at an imaginary spot where his bugging device might be (since he does not know its real location)."[74]

Between the two extremes of avoiding conversation and speaking directly into the presumed microphones was the most commonly reported type of behaviour: the taking of modest precautions.

> Before entering the apartment, [refusenik Yosef T.] warned us not to speak in the passage ways, we entered the apartment, he quickly put on the radio loudly and closed all of the window shades.
>
> (Dennis Freilich, 1973)[75]

> Mrs R. thought that the apartment was bugged through the telephone, and she turned the dial and inserted a pencil in it.
>
> (Joel Sprayregen, 1970)[76]

> Pavel unplugs his phone as soon as we enter his apartment.
>
> (Victor Borden, 1985)[77]

Travelogues rarely evaluated the efficacy of these efforts. They simply noted them. But in noticing and writing about these small attempts to protect the boundaries that made domestic space private, travel writers invoked the authority of the locals to represent the Soviet state as possessing a penetrating panoptic power. Such a view accorded with their own preconceptions, as evidenced by thousands of separate accounts of travellers' nervous preparations for hiding their meetings from customs officials, Intourist guides, and others. But observing how refuseniks took precautions of their own served in the travel narratives to validate these apprehensions by suggesting that Soviet Jews shared them too, even in – especially in – their own homes.

Conclusion

From the mid-1950s onwards, and with increasing momentum with detente, Westerners used tourism as a way of engaging personally with the Cold War. This was not simply tourism in the Cold War era, or tourism in a Cold War context, but tourism *of* the Cold War, positioning the Cold War itself as a tourist destination. Fundamental to the Western construction of the Cold War as a tourist attraction was the notion of the Soviet Union as a closed society, hidden and inaccessible behind its Iron Curtain. Its spaces, forbidden and foreboding, were for that very reason alluring. The promise of tourism was the promise of penetrating these spaces to see behind the Iron Curtain, as it were. It was, of course, a chimera. Cold War frontiers were omnipresent, and not coterminous with the geopolitical border. The fact of entering the Soviet Union did not render all its spaces accessible. On the contrary, government control over visitors' itineraries constrained Westerners' ability to see much beyond the public spaces that were open to viewing.[78] In this sense, tourism merely moved the frontier, pushing back the realm of the forbidden. Domestic space remained largely inaccessible to most travellers, as did the unstructured contact with locals that such spaces afforded. In the symbolic economy of Cold War tourism, home visits, unofficial and unsanctioned, were the ultimate "off the beaten track," and were valued accordingly.

Human rights tourism is notable in the annals of Cold War culture as one of the few modes of travel that regularly opened Soviet domestic space to a Western gaze. Organisations working to enable Soviet Jewish emigration provided thousands of tourists an opportunity to tour the Cold War from an unusual vantage point. From inside refuseniks' apartments, human rights tourists tried to make sense of the USSR by examining how such homes stood in relation to the society at large. Their writings reveal that they tended to see this domestic space both as an antithesis to Soviet space – a warm and hospitable haven in the midst of a cold and hostile land – and as a Soviet space in its own right – a socialist creation penetrated by a police state. These representations took negative Western images of the Soviet Union and re-energised them, imbuing them with all the symbolic power associated with the tourist impulse to go behind the scenes. "There are bars on the one window in the small living room," Victor Borden wrote of a refusenik's apartment in Leningrad. "It truly reminds me of the prison that it is."[79]

It was one of the ironies of the Cold War that detente itself paved the way for this, widening the tourist channels, which human rights groups seized on to advance their cause. The encounters that took place in refuseniks' apartments fostered little tolerance for, much less acceptance of, the Soviet system. Rather, they reinforced transnational solidarity among those committed to challenging Soviet power from both sides of the Iron Curtain.

Notes

1   The author gratefully acknowledges the support of the National Endowment for the Humanities (FT-229663-15), the University of Michigan's Frankel Institute for Advanced Judaic Studies, the Brandeis-Genesis Institute for Russian Jewry, the Hadassah-Brandeis Institute, the Western Reserve Historical Society, and the Robert Penn Warren

Center for the Humanities at Vanderbilt University. I wish to thank the American Jewish Historical Society, Allison Schachter, Holly McCammon, Seth Jacobson, and my research assistants, Elizabeth Dultz, Roxana Maria-Aras, Andrea Becker, and Katherine Pullen.

Marge Gordon, "Trip Report of Marge & Bob Gordon," 6–20 July 1975, 3, RG I-487, Box 90, Folder 8, Records of Action for Soviet Jewry, Archive of the American Soviet Jewry Movement (hereafter AASJM), American Jewish Historical Society, New York (hereafter AJHS).

2 Gordon, "Trip Report," 1. On the "gray and unsmiling Moscow" trope in post-war American travel writing about the Soviet Union, see Zachary Jonathan Jacobson, "American Studies, the Soviet Union: A Cultural History of US – Soviet Encounters through the Cold War" (PhD diss., Northwestern University, 2014), ch. 1.

3 Gordon, "Trip Report," 1.

4 Over four days in Moscow, they visited two families once; two, twice; and two, three times. In Kiev, they visited one family at home twice in two days. In Leningrad, they reported only two home visits over four days, fewer than they were hoping for. Of this, Gordon wrote "Bob asked Zhanna several times to arrange a meeting with other [Jewish activists] in Leningrad, but because of the animosity among the different factions, Zhanna tried to discourage any other meetings" (Gordon, "Trip Report," 4).

5 Several archives in the US, Western Europe and Israel now hold travelogues filed with Soviet Jewry movement organizations. The bulk of those that are publicly available are held by the American Jewish Historical Society's Archive of the American Soviet Jewry Movement in New York, of which over 2,000 are available online, at www.ajhs. org/digital-collections. Reports by some 6,000 travellers filed with the Israeli government's Soviet Jewry agency, Nativ, remain classified, but were the primary source for Shelomoh Rozner, *Bi-netiv ha-demamah: ha-pe'ilut ha-ḥashai'it lema'an Yehude Berit ha-Mo'atsot* [Silent Route: The Clandestine Support for Soviet Jews], ed. Eli Somer and Tzemach Jacobson (Jerusalem: Zalman Shazar Center, 2012), an in-house history of Nativ's use of tourism.

6 Gordon, "Trip Report," 4. I have followed the convention of referring to their Soviet Jewish hosts by first name and last initial, except when they were public figures who featured prominently in the Western press.

7 Jonathan Culler, "Semiotics of Tourism," *American Journal of Semiotics* 1, no. 1/2 (1981): 127–40; Dean MacCannell, *The Tourist: A New Theory of the Leisure Class* (New York: Schocken, 1989); John Urry, *The Tourist Gaze* (London: SAGE, 1990).

8 Kay Thompson, *Eloise Goes to Moscow*, ill. Hilary Knight (New York: Simon & Schuster, 1959), 40–41, the fourth in the popular *Eloise* series about the adventures of a precocious little girl who lives on the top floor of New York's Plaza Hotel, has a full-spread illustration of an endless queue to visit Lenin's tomb. Visiting the Soviet Union, Eloise proclaims, "They stand in line for everything in Moscow."

9 Gordon, "Trip Report," 4.

10 The rhetoric of "peace through tourism" was not new. First popularized in the interwar years, it had become the refrain of national tourist offices in Western Europe in the era of post-war reconstruction, and was later picked up by the UN, which celebrated 1967 as International Tourist Year under the slogan "Tourism, Passport to Peace." Sune Bechmann Pedersen, "Peace through Tourism: A Brief History of a Popular Catchphrase," in *Cultural Borders and European Integration*, ed. Mats Andrén (Gothenburg: Centrum för Europaforskning vid Göteborgs Universitet, 2017), 29–37.

11 Eugene Fodor and Robert C. Fisher, eds., *Fodor's Soviet Union 1974–1975* (New York: David McKay, 1974), 6.

12 Shawn Connelly Salmon, "To the Land of the Future: A History of Intourist and Travel to the Soviet Union, 1929–1991" (PhD diss., University of California, Berkeley, 2008), 287.

13 Jason Paur, "Oct. 4, 1958: 'Comets' Debut Trans-Atlantic Jet Age," *Wired*, 4 October 2010. www.wired.com/2010/10/1004first-transatlantic-jet-service-boac; Farnsworth

Fowle, "Aeroflot Leads on Moscow Run, But Pan Am is Satisfied," *New York Times*, 7 July 1969, p. 66.

14 Soviet figures put the number of US citizens entering in 1964 at 23,000; the 1970 and 1975 figures were for American and Canadian travellers combined. Douglas W. Cray, "Communist Lands Wooing Tourists," *New York Times*, 3 January 1965, section 3, p. 5; US Bureau of the Census, *USA/USSR: Facts and Figures* (Washington: US Government Printing Office, 1991), 2-11-2-12, tables 2.12, 2.13. Using data from the UN, others have calculated Soviet tourism to NATO countries to have risen from 11,415 in 1958 to 41,349 in 1977, and NATO tourism to the USSR to have risen from 62,793 in 1958 to 442,929 in 1977. Randolph M. Siverson, Alexander J. Groth, and Marc Blumberg, "Soviet Tourism and Détente: 1958–1977," *Studies in Comparative Communism* 13, no. 4 (1980): 365–66.

15 Anne E. Gorsuch, *All This is Your World: Soviet Tourism at Home and Abroad After Stalin* (Oxford: Oxford University Press, 2011), 103–29; Robert A. Hornsby, "The Enemy Within? The Komsomol and Foreign Youth Inside the Post-Stalin Soviet Union, 1957–1985," *Past & Present* 232, no. 1 (2016): 237–78; Alexander Hazanov, "Porous Empire: Foreign Visitors and the Post-Stalin Soviet State" (PhD diss., University of Pennsylvania, 2016); Vardan Bagdasaryan, Igor Orlov, Iosif Schneider, Alexander Fedulin, and Konstantin Mazin, eds., *Sovetskoe zazerkal'e: innostrannyi turizm v SSSR v 1930–1980 gody* [The Soviet Looking Glass: Foreign Tourism to the USSR, 1930s–1980s] (Moscow: Forum, 2007), 241–52.

16 Gorsuch, *All This is Your World*, 189–90. Alexei Yurchak, *Everything Was Forever, Until It Was No More: The Last Soviet Generation* (Princeton: Princeton University Press, 2006). On counter-effects that deepened Soviet institutional commitments to authoritarianism, see Hazanov, "Porous Empire."

17 Yale Richmond, *Cultural Exchange and the Cold War: Raising the Iron Curtain* (University Park, PA: Penn State University Press, 2003).

18 For the contribution of human rights activism to the collapse of communism and the end of the Cold War, see Daniel Charles Thomas, *The Helsinki Effect: International Norms, Human Rights, and the Demise of Communism* (Princeton: Princeton University Press, 2001); Sarah B. Snyder, *Human Rights Activism and the End of the Cold War: A Transnational History of the Helsinki Network* (New York: Cambridge University Press, 2011).

19 For the Soviet Jews' activism, see Yaacov Ro'i, ed., *The Jewish National Movement in the Soviet Union* (Washington, DC: Woodrow Wilson Center Press, 2012); on Israeli efforts, see Nehemiah Levanon, *Ha-kod 'Nativ'* [Code Name 'Nativ'] (Tel Aviv: Am Oved, 1995).

20 Khvorostianyi to CPSU Central Committee, memo, 2 September 1959, in Boris Morozov, *Documents on Soviet Jewish Emigration* (Portland: Frank Cass, 1999), 40–41.

21 Of the 450 synagogues in the Soviet Union in 1956, 392 (87 per cent) were closed by 1972. In Russia, Ukraine, and the Baltic States, where the majority of Soviet Jews lived, there was one synagogue per 62,500 Jews, compared to one church per 2,000 Russian Orthodox Christians. William Korey, *The Soviet Cage* (New York: Viking, 1973), 44–45.

22 Exposing the multifaceted character of Soviet state-sponsored anti-Semitism was a key task of the campaigns in the West, with numerous books and articles on the topic. The best known was Elie Wiesel, *The Jews of Silence* (New York: Holt, Rinehart & Winston, 1966).

23 General histories of the campaign to free Soviet Jews include William W. Orbach, *The American Movement to Aid Soviet Jews* (Amherst: University of Massachusetts Press, 1979); Yaacov Ro'i, *The Struggle for Soviet Jewish Emigration, 1948–1967* (Cambridge: Cambridge University Press, 2003); Henry L. Feingold, *Silent No More: Saving The Jews of Russia, The American Jewish Effort, 1967–1989* (Syracuse:

Syracuse University Press, 2004); Pauline Peretz, *Le combat pour les Juifs sovié-tiques: Washington – Moscou – Jérusalem, 1953–1989* (Paris: Armand Colin, 2006); and Gal Beckerman, *When They Come for Us We'll Be Gone: The Epic Struggle to Save Soviet Jewry* (New York: Houghton Mifflin Harcourt, 2010). The largest rally for Soviet Jewry brought 250,000–300,000 people to the National Mall in Washington, DC, on 6 December 1987, the eve of a US–Soviet summit in the US capital.

24 The US Congress, for example, passed the Jackson–Vanik amendment to the US–Soviet trade bill of 1975, which made trade preferences for the Soviet Union conditional that it respect the right to free emigration. For its part, the Netherlands provided consular services in Moscow to Soviet Jews on Israel's behalf after the USSR broke diplomatic ties with Israel in 1967. For over half a million Soviet Jews, the path to emigration passed through the Dutch embassy. For Jackson–Vanik, see Fred A. Lazin, *The Struggle for Soviet Jewry in American Politics: Israel versus the American Jewish Establishment* (Lanham: Lexington, 2005). For the work of the Dutch government, see Petrus Buwalda, *They Did Not Dwell Alone: Jewish Emigration from the Soviet Union, 1967–1990* (Washington, DC: Woodrow Wilson Center, 1997).

25 Levanon, *Ha-kod 'Nativ'*; Peretz, *Le combat pour les Juifs soviétiques*, 62–76.

26 Rozner, *Bi-netiv ha-demamah*, 37–42, 46.

27 Rabbinic organizations had been sending official delegations to the Soviet Union since 1956. Western Soviet Jewry organizations had relied on tourists for information about Soviet Jewry throughout the 1960s, but on an ad hoc basis. Their systematic mobilization of tourism began only in the early 1970s. David Hollander, Herschel Schacter, Samuel Adelman, Emanuel Rackman, and Gilbert Klaperman, Untitled trip report of Rabbinical Council of America delegation to Soviet Union, Romania, Czechoslovakia and Poland, 21 June–2 August 1956, MS – 763, Series H, Subseries 4, Box 49, Folder 3, Rabbi Herbert A. Friedman Collection, Jacob Rader Marcus Center, American Jewish Archives, Cincinnati, Ohio. For the Union of Councils, see Shaul Kelner, "People-to-People: Cleveland's Jewish Community and the Exodus of Soviet Jews," in *Cleveland Jews and the Making of a Midwestern Community*, ed. John Grabowski and Sean Martin (New Brunswick: Rutgers University Press, 2020); Hester Beckman to Chapter and Division Presidents, Re: "Briefing Kits for Tourists to Russia," memo, 20 January 1972, RG I-77, Box 48, Folder 6, Records of the American Jewish Congress, AASJM, AJHS.

28 The overwhelming majority of American Jews at the time traced their ancestry to Yiddish-speaking parents and grandparents who had emigrated from Tsarist Russia. The refuseniks' parents and grandparents were, by and large, the emigrants' Yiddish-speaking relatives who remained in Russia. Guests and hosts were thus two branches of a single ethno-linguistic community, divided by emigration, one branch of which exchanged Yiddish for American English, the other for Soviet Russian.

29 Hebrew language studies were part of the cultural resistance secretly practised by Soviet Jewish activists preparing to emigrate to Israel. Ari Volvovsky, "The Teaching and Study of Hebrew," in *The Jewish National Movement in the Soviet Union*, ed. Yaacov Ro'i (Washington, DC: Woodrow Wilson Center, 2012), 334–55; Vera Yedidya, "The Struggle for the Study of Hebrew," in *Jewish Culture and Identity in the Soviet Union*, ed. Yaacov Ro'i and Avi Becker (New York: New York University Press, 1991), 136–67. Western Jews who had some facility in spoken Hebrew tended to come by it in ethnic studies classes at university, clerical training in rabbinical seminaries, or language immersion courses (*ulpanim*) in Israel. Hebrew was also the native tongue of Israeli-born travellers. Unlike the American and European NGOs, Nativ emissaries in Western Europe and North America made a point of recruiting Hebrew-speaking travellers (Rozner, *Bi-netiv ha-demamah*, 71, 74).

30 See, for example, Phil Baum and Zev Furst, "How to Find and Meet Soviet Jews: Briefing Kit for Travelers to the U.S.S.R." (New York: American Jewish Committee,

1972), RG I-181A, Box 91, Folder 2, Records of the National Conference on Soviet Jewry (hereafter cited as NCSJ I-181A), AASJM, AJHS.

31 Such gift exchanges were similar to those between American journalists and Soviet dissidents, described in Barbara Walker, "The Moscow Correspondents, Soviet Human Rights Activists, and the Problem of the Western Gift," in *Americans Experience Russia: Encountering the Enigma, 1917 to the Present*, ed. Choi Chatterjee and Beth Holmgren (New York: Routledge, 2013), 139–57.

32 This chapter is based on a grounded-theory content analysis of the travelogues in the AASJM and other repositories (coded in the ATLAS–TI qualitative data analysis software package), and on a statistical analysis of the travelogues, using a database created by the author with the help of research assistants.

33 Nat Kameny, "Our Soviet Journey," 1–9 March 1974, 7, NCSJ I-181A, Box 64, Folder 16, AASJM, AJHS. A revised version was published the following month in the newsletter of the Anti-Defamation League of B'nai B'rith. Nat Kameny, "Nine Days in the Soviet Union," *ADL Bulletin* 31, no. 4 (April 1974), 4–5, 8.

34 Descriptions of arrests, interrogations and expulsions can be found, for example, in Caroline Rabinowitz and Ryan Hileman, "Trip Report," 3–21 April 1986, 23–29, RG I-507, Box 9, Folder 7, Records of Seattle Action for Soviet Jewry, AASJM, AJHS; and Joel Sandberg and Adele Sandberg, "Trip to the U.S.S.R.," 6–14 May 1975, 4–7, RG P-906, Box 90, Folder 1, Papers of Morey Schapira [hereafter Schapira Papers P-906], AASJM, AJHS. An example of Soviet press articles denouncing Jewish travellers by name is V. Nilov, "*Oplata . . . galetami,*" ["Payment . . . with biscuits"] *Izvestia*, 9 April 1976, with English translation in RG I-540, Box 9, Folder 7, Records of the Washington Committee for Soviet Jewry [hereafter WCSJ I-540], AASJM, AJHS.

35 Sarolta Klenjanszky, "Touristic Relations between Eastern and Western Europe through the Lens of the City Twinning Movement (1960–1990)," paper presented at Crossing the Iron Curtain: Tourism and Travelling in the Cold War, Universiteit van Amsterdam, Amsterdam, 7 April 2017.

36 Hornsby, "The Enemy Within?"

37 Salmon, "To the Land of the Future," 259.

38 Jacobson, "American Studies," 33; Walker, "The Moscow Correspondents," 140.

39 Clergy and professionals in Jewish community organizations were more likely to travel specifically to contact Soviet Jews. Other than this, there was no substantial difference in the occupational profile of the volunteer activists travelling on behalf of the movement and casual tourists swept up in the cause – both were drawn from the ranks of educated, affluent, and mobile American Jews.

40 Most activists were seeking not merely to emigrate, but to emigrate to Israel. When the Soviet Union first allowed Jewish emigration in the 1970s, hundreds of thousands of non-activist Jews left the country, mostly for the US, Canada, and Germany. In the late 1980s, faced with the largest Soviet Jewish exodus, the US, Soviet, and Israeli governments agreed to channel the majority to Israel. See Lazin, *The Struggle for Soviet Jewry*.

41 These numbers are illustrative, not definitive. Beyond the thousands of travellers' records in the AASJM, there are thousands that are not, including most of the 6,000 sent by Israel's Nativ and the majority of those by Western European travellers. The European activists' modus operandi of choice was a weekend excursion to Leningrad from a Scandinavian hop-off point, so the 85 per cent figure reported here for Leningrad is probably low. Rozner, *Bi-netiv ha-demamah*, 62–63, 78.

42 My reading of these travelogues is informed by analytical strategies offered in Carl Thompson, *Travel Writing* (New York: Routledge, 2011).

43 In reality, the imperative derives from the hope of escaping the semiotic readings of place – which by definition cannot be realized in the context of tourism, since to sightsee is to read places as symbols. See Culler, "Semiotics of Tourism"; Shaul Kelner, *Tours That Bind: Diaspora, Pilgrimage and Israeli Birthright Tourism* (New York: New York University Press, 2010), 140.

44 MacCannell, *The Tourist*, 93–94, emphasis added.
45 It is important to recognize the problematics of Western notions of public and private when talking of the Soviet context. Addressing whether it is possible to speak of Soviet homes as "private" spaces, Susan E. Reid, "The Meaning of Home: 'The Only Bit of the World You Can Have to Yourself,'" in *Borders of Socialism: Private Spheres of Soviet Russia,* ed. Lewis H. Siegelbaum (New York: Palgrave MacMillan, 2006), 145, 148 argues that if the "public" is that which is visible and collective, Soviet homes functioned as private space not because they offered full concealment or solitude, but because they offered some measure of control over the boundary with the public. Home was where Soviet citizens enjoyed "discretion over disclosure of information about oneself, the right to make decisions, to promulgate rules of action, to dispose over resources and space, and to choose association with others." For a brief overview of scholarly approaches to the question of the "private" in the late Soviet context, see Ekaterina Emeliantseva, "The Privilege of Seclusion: Consumption Strategies in the Closed City of Severodvinsk," *Ab Imperio* 2011, no. 2 (2011): 238–59. For an analysis from the period when the exodus of Soviet Jewry was taking place, see Vladimir Shlapentokh, *Public and Private Life of the Soviet People: Changing Values in Post-Stalin Russia* (New York: Oxford University Press, 1989), 164–89 & *passim*.
46 Gordon, "Trip Report," 1.
47 Ruth Nordlicht, "A Personal Glimpse from Behind the Iron Curtain," 6 July 1975, NCSJ I-181A, Box 69, Folder 17, AASJM, AJHS.
48 Harriet Goldberg, "Trip to the Soviet Union," 9 January 1980, 3, NCSJ I-181A, Box 61, Folder 2, AASJM, AJHS.
49 Victor Borden, "July 1985 USSR Mission," 4–13 July 1985, 5, NCSJ I-181A, Box 56, Folder 10, AASJM, AJHS.
50 Gordon, "Trip Report," 3.
51 Goldberg, "Trip to the Soviet Union," 12–13.
52 National Conference on Soviet Jewry, "Debriefing Report," 1988, NCSJ I-181A, Box 53, Folder 4, AASJM, AJHS.
53 Similarly, while tourists were asked to photograph refuseniks, they were not instructed to photograph their homes per se. As one Union of Councils for Soviet Jews guide for their briefers had it, "Briefing Procedure," 10, 1989, WCSJ I-540, Box 11, Folder 9, AASJM, AJHS: "It's important for [the] tourist to take pictures of Refuseniks, especially close ups. Although [the] tourist will naturally want pictures of himself with the Refuseniks, remind him to take some of the Refuseniks by themselves." Many of the pictures did, nevertheless, reveal something about the settings.
54 Activists' use of narrative writing (journaling) as a way of developing self-understanding for their cause and their participation in it (irrespective of whether they were writing with publication for a broader audience in mind) is yet another use of storytelling by social movements beyond those identified in Francesca Polletta, *It Was Like a Fever: Storytelling in Protest and Politics* (Chicago: University of Chicago Press, 2006).
55 Douglas Adams, *The Long Dark Tea-Time of the Soul* (New York: Gallery, 2014 [1988]), 146–47.
56 The size of larger apartments was often noted as exceptional.
57 Gordon, "Trip Report," 2.
58 Trudy Shecter and Debbie Shecter, "Visit to the Soviet Union," 10–20 July 1975, 2, Schapira Papers P-906, Box 99, Folder 1, AASJM, AJHS.
59 Rochelle Ginsburg, Untitled trip report, July 1986, 11–12, NCSJ I-181A, Box 60, Folder 33, AASJM, AJHS.
60 Dennis Freilich and Estelle Freilich, "Report of Trip to the Soviet Union," 3–17 June 1973, 5, NCSJ I-181A, Box 60, Folder 4, AASJM, AJHS. Evidence in the text indicates that Dennis authored the sections of the report quoted here.
61 Although Nixon won the election with 61 per cent of the popular vote to McGovern's 38 per cent, American Jews voted overwhelmingly for the latter. McGovern

garnered 65 per cent of the American Jewish vote; Nixon 35 per cent. Moreover, American Jewish activists in the campaign for Soviet Jewry opposed a normalization of US–Soviet relations that ignored Soviet human rights abuses (hence the Jackson–Vanik Amendment, linking American trade preferences to Soviet respect for free emigration). Nixon viewed this as a threat to detente. The opposing viewpoints set the stage for conflict between the movement and the administration. For the conflict over Jackson–Vanik, see Lazin, *The Struggle for Soviet Jewry*; for Nixon's uncensored views on American Jewish activism, see Elspeth Reeve, "Some Newly Uncovered Nixon Comments on the Subjects of Jews and Black People," *The Atlantic*, 21 August 2013. www.theatlantic.com/politics/archive/2013/08/some-new-comments-richard-nixon-subject-jews-and-blacks/311870/.

62 Jerry Lewis, "Report of Visit to Moscow," 9–11 December 1973, 2, NCSJ I-181A, Box 67, Folder 15, AASJM, AJHS.
63 Borden, "July 1985 USSR Mission," 2.
64 In Jewish Sabbath ritual, before blessings are recited over challah bread, the loaf is usually covered with a cloth. Decorative challah covers are a common form of Judaica. The Bordens gave such a cover to the Ioffes.
65 Borden, "July 1985 USSR Mission," 7–8.
66 Ibid., 8–9.
67 Ibid., 15, emphasis added.
68 Ibid., 15, emphasis added.
69 Ibid., 25, emphasis added.
70 Ibid., 13–14.
71 Kameny, "Nine Days in the Soviet Union," 5. The draft version was more colourful: "Arriving at Professor Lerner's apartment, brushing past two witch-like spies hovering about the lobby" (Kameny, "Our Soviet Journey," 3).
72 Jules Lippert, Untitled trip report, 27 July – 1 August 1973, 2, NCSJ I-181A, Box 67, Folder 23, AASJM, AJHS. The term "plumbers" was laden with meaning in 1973, as the well-known moniker for the perpetrators of the Watergate break-in.
73 Ahaviah Scheindlin and Bill Aron, "Trip to U.S.S.R.," 5–23 October 1981, 12, NCSJ I-181A, Box 71, Folder 23, AASJM, AJHS.
74 Gordon, "Trip Report," 2.
75 Freilich and Freilich, "Report of Trip to the Soviet Union," 5.
76 Joel Sprayregen, "Memorandum on Contacts with Jews in the Soviet Union," 4 June 1970, 5, NCSJ I-181A, Box 74, Folder 16, AASJM, AJHS.
77 Borden, "July 1985 USSR Mission," 12.
78 For controls that restricted tourists' mobility in the USSR, even in the midst of Soviet efforts to structure official people-to-people contacts, see Hornsby, "The Enemy Within?"; Salmon, "To the Land of the Future," 173–74, 258–60.
79 Borden, "July 1985 USSR Mission," 21.

References

Adams, Douglas. *The Long Dark Tea-Time of the Soul*. New York: Gallery, 2014 [1988].
Bagdasaryan, Vardan, Igor Orlov, Iosif Schneider, Alexander Fedulin, and Konstantin Mazin, eds. *Sovetskoe zazerkal'e: Innostrannyi turizm v SSSR v 1930–1980-e gody* [The Soviet Looking Glass: Foreign Tourism in the USSR, 1930s – 1980s]. Moscow: Forum, 2007.
Bechmann Pedersen, Sune. "Peace through Tourism: A Brief History of a Popular Catch-phrase." In *Cultural Borders and European Integration*, edited by Mats Andrén, 29–37. Gothenburg: Centrum för Europaforskning vid Göteborgs universitet, 2017.
Beckerman, Gal. *When They Come for Us We'll Be Gone: The Epic Struggle to Save Soviet Jewry*. New York: Houghton Mifflin Harcourt, 2010.

Buwalda, Petrus. *They Did Not Dwell Alone: Jewish Emigration from the Soviet Union, 1967–1990*. Washington, DC: Woodrow Wilson Center Press, 1997.

Culler, Jonathan. "Semiotics of Tourism." *American Journal of Semiotics* 1, no. 1/2 (1981): 127–40.

Emeliantseva, Ekaterina. "The Privilege of Seclusion: Consumption Strategies in the Closed City of Severodvinsk." *Ab Imperio* 2011, no. 2 (2011): 238–59.

Feingold, Henry L. *Silent No More: Saving the Jews of Russia, The American Jewish Effort, 1967–1989*. Syracuse: Syracuse University Press, 2004.

Fodor, Eugene and Robert C. Fisher, eds. *Fodor's Soviet Union 1974–1975*. New York: David McKay, 1974.

Fowle, Farnsworth. "Aeroflot Leads on Moscow Run, But Pan Am Is Satisfied." *New York Times*, 7 July 1969, p. 66.

Gorsuch, Anne E. *All This Is Your World: Soviet Tourism at Home and Abroad After Stalin*. Oxford: Oxford University Press, 2011.

Hazanov, Alexander. "Porous Empire: Foreign Visitors and The Post-Stalin Soviet State." PhD diss., University of Pennsylvania, 2016.

Hornsby, Robert A, "The Enemy Within? The Komsomol and Foreign Youth inside the Post-Stalin Soviet Union, 1957–1985." *Past & Present* 232, no. 1 (2016): 237–78.

Jacobson, Zachary Jonathan. "American Studies, the Soviet Union: A Cultural History of US – Soviet Encounters through the Cold War." PhD diss., Northwestern University, 2014.

Klenjanszky, Sarolta. "Touristic Relations between Eastern and Western Europe through the Lens of the City Twinning Movement (1960–1990)." Paper presented at the conference Crossing the Iron Curtain: Tourism and Travelling in the Cold War, Universiteit van Amsterdam, Amsterdam, 7 April 2017.

Korey, William. *The Soviet Cage*. New York: Viking Press, 1973.

Lazin, Fred A. *The Struggle for Soviet Jewry in American Politics: Israel versus the American Jewish Establishment*. Lanham: Lexington, 2005.

Levanon, Nehemiah. *Ha-kod 'Nativ'* [Code Name 'Nativ']. Tel Aviv: Am Oved, 1995.

MacCannell, Dean. *The Tourist: A New Theory of the Leisure Class*. New York: Schocken, 1989.

Morozov, Boris, ed. *Documents on Soviet Jewish Emigration*. Portland: Frank Cass, 1999.

Orbach, William W. *The American Movement to Aid Soviet Jews*. Amherst: University of Massachusetts Press, 1979.

Paur, Jason. "Oct. 4, 1958: 'Comets' Debut Trans-Atlantic Jet Age." *Wired*, 4 October 2010. www.wired.com/2010/10/1004first-transatlantic-jet-service-boac.

Peretz, Pauline. *Le combat pour les Juifs soviétiques: Washington – Moscou – Jérusalem, 1953–1989*. Paris: Armand Colin, 2006.

Polletta, Francesca. *It was Like a Fever: Storytelling in Protest and Politics*. Chicago: University of Chicago Press, 2006.

Reeve, Elspeth. "Some Newly Uncovered Nixon Comments on the Subjects of Jews and Black People." *The Atlantic*, 21 August 2013. www.theatlantic.com/politics/archive/2013/08/some-new-comments-richard-nixon-subject-jews-and-blacks/311870/.

Reid, Susan E. "The Meaning of Home: 'The Only Bit of the World You Can Have to Yourself.'" In *Borders of Socialism: Private Spheres of Soviet Russia*, edited by Lewis H. Siegelbaum, 145–70. New York: Palgrave MacMillan, 2006.

Richmond, Yale. *Cultural Exchange and the Cold War: Raising the Iron Curtain*. University Park, PA: Penn State University Press, 2003.

Ro'i, Yaacov. *The Struggle for Soviet Jewish Emigration, 1948–1967*. Cambridge: Cambridge University Press, 2003.

———, ed. *The Jewish National Movement in the Soviet Union.* Washington, DC: Woodrow Wilson Center Press, 2012.

Rozner, Shelomoh. *Bi-netiv ha-demamah: ha-pe'ilut ha-ḥashai'it lema'an Yehude Berit ha-Mo'atsot* [Silent Route: The Clandestine Support for Soviet Jews], edited by Eli Somer and Tzemach Jacobson. Jerusalem: Zalman Shazar Center, 2012.

Salmon, Shawn Connelly. "To the Land of the Future: A History of Intourist and Travel to the Soviet Union, 1929–1991." PhD diss., University of California, Berkeley, 2008.

Shaul Kelner. "People-to-People: Cleveland's Jewish Community and the Exodus of Soviet Jews." In *Cleveland Jews and the Making of a Midwestern Community*, edited by John Grabowski and Sean Martin. New Brunswick: Rutgers University Press, 2020.

———. *Tours That Bind: Diaspora, Pilgrimage and Israeli Birthright Tourism.* New York: New York University Press, 2010.

Shlapentokh, Vladimir. *Public and Private Life of the Soviet People: Changing Values in Post-Stalin Russia.* New York: Oxford University Press, 1989.

Siverson, Randolph M., Alexander J. Groth, and Marc Blumberg. "Soviet Tourism and Détente: 1958–1977." *Studies in Comparative Communism* 13, no. 4 (1980): 356–68.

Snyder, Sarah B. *Human Rights Activism and the End of the Cold War: A Transnational History of the Helsinki Network.* New York: Cambridge University Press, 2011.

Thomas, Daniel Charles. *The Helsinki Effect: International Norms, Human Rights, and the Demise of Communism.* Princeton: Princeton University Press, 2001.

Thompson, Carl. *Travel Writing.* New York: Routledge, 2011.

US Bureau of the Census, *USA/USSR: Facts and Figures.* Washington: US Government Printing Office, 1991.

Urry, John. *The Tourist Gaze.* London: SAGE, 1990.

Volvovsky, Ari. "The Teaching and Study of Hebrew." In *The Jewish National Movement in the Soviet Union*, edited by Yaacov Ro'i, 334–55. Washington, DC: Woodrow Wilson Center Press, 2012.

Walker, Barbara. "The Moscow Correspondents, Soviet Human Rights Activists, and the Problem of the Western Gift." In *Americans Experience Russia: Encountering the Enigma, 1917 to the Present*, edited by Choi Chatterjee and Beth Holmgren, 139–57. New York: Routledge, 2013.

Wiesel, Elie. *The Jews of Silence.* New York: Holt, Rinehart & Winston, 1966.

Yedidya, Vera. "The Struggle for the Study of Hebrew." In *Jewish Culture and Identity in the Soviet Union*, edited by Yaacov Ro'i and Avi Becker, 136–67. New York: New York University Press, 1991.

Yurchak, Alexei. *Everything Was Forever, Until It Was No More: The Last Soviet Generation.* Princeton: Princeton University Press, 2006.

6 "Much more freedom of thought than expected there"

Rosey E. Pool, a Dutch fellow traveller on holiday in the Soviet Union (1965)

Lonneke Geerlings

"Much more freedom of thought than expected there," the Dutch translator Rosey E. Pool (1905–1971) wrote to a friend. It was late 1965 and she had just travelled across the Soviet Union, together with her life partner, Ursel "Isa" Isenburg (1901–1987). The trip provided a "most interesting experience from many angles," Pool wrote, perhaps hinting at their multilayered personalities that made them receptive to different perspectives on marginalised communities.[1]

Pool and Isenburg were a lesbian couple; both were Jewish; both were followers of the Baha'i faith;[2] and both were experienced world travellers. Moreover, Pool's experience in interwar Popular Front movements and their shared anti-racist activism greatly influenced their conception of, encounters with, and interpretations of Soviet society. This chapter provides an individual, biographical approach to Soviet travel during the Cold War. Their journeys and their travel writing reveal a heterogeneous message about the Soviet Union, which becomes clear through a close reading. Fellow travellers used positive remarks about the Soviet Union to affirm political sympathies, while the careful mention of public figures was meant to show which side of the political spectrum they were on – all applied in such a way as to maintain plausible deniability. This chapter actively reads their writings and correspondence "against the grain" to detect subtle meanings.

Rosey E. Pool was not only a translator, educator, and anthologist of African-American poetry. She was also an expatriate, a migrant, and at times, a tourist. Her transnational life brought her to various hotbeds of history: Berlin, as the Weimar Republic became Nazi Germany; Amsterdam during the war, where she taught Anne Frank before she went into hiding; a Nazi transit camp, from which she miraculously escaped; Mississippi in the midst of the civil rights movement. In the late 1940s she moved to London, sharing a home with Isenburg, a German Jewish radiologist who had fled Nazi Germany in the 1930s. Isenburg and Pool knew each other from their Berlin days in the 1920s and now they reconnected. Until Pool's death in 1971 they spent the rest of their lives together – at home and on the road for over 23 years. To the outside world they were inseparable friends. Pool referred to her in letters as "my dear friend," her "flatmate" or "my living and business partner." To close friends, however, it was obvious that they were a couple.

In the 1950s and 1960s the two travelled extensively. For a short time they even ran their own travel agency in London, "Allways [*sic*] Travel Service."[3] Although they closed the agency after a few years, the two women continued to "allways" travel. As travel experts they now organised and booked their own trips and travelled independently around the globe. Their holidays included Israel, Greece (1954), Mexico, France (1960), Norway (1961), the West Indies, Haiti, Jamaica, Canada (1963), Greece, Yugoslavia, North Africa, Italy (1964), Latin America (1967), and Turkey (1968). Moreover, Pool also travelled on her own to the US (as a Fulbright scholar in 1959–60, and to spend several semesters as a visiting scholar in the Southern US between 1962 and 1967), and attended festivals and conferences in Nigeria, Ghana (1961), and Senegal (1966). Because Pool spoke several languages she thought of "many parts of the world as a kind of home," seeing herself as a sort of "gypsy."[4] Isenburg, Prussian by birth, was not a polyglot like Pool, but nonetheless shared her passion for travel. She attended international radiology conferences and frequently visited Paris, which she had "adopted" as her hometown of choice. "Isa never feels a holiday is complete without her beloved Paris," Pool wrote.[5] As individual travellers who moved between

Figure 6.1 Pool (right) and Isenburg on the balcony of their home in London in the 1960s

Source: © Jewish Historical Museum, Amsterdam

countries, cultures, and contexts, they experienced unhomeliness and ambiguity of place, and their sense of belonging and identity could be transitional.

Their 1965 holiday to the Soviet Union was part of an increased interest in travelling behind the Iron Curtain. Khrushchev's Thaw saw the lifting of travel restrictions, which enabled Western tourists and students to visit the Soviet Union in great numbers.[6] It was often organised by Soviet friendship organisations that had emerged in the West in previous decades. Vernu-reizen, the travel agency of the Dutch friendship organisation *Vereniging Nederland-USSR* (1947–1997), claimed that in 1963 no fewer than 800,000 foreigners visited the Soviet Union for a holiday – mostly Americans, but also some from the Netherlands. "Travel behind the Iron Curtain is 'hot,'" one Dutch newspaper reported.[7] It neverthe-less remained an unusual travel destination. Vernu-reizen assured the Amsterdam newspaper *Het Parool* that visiting the Soviet Union was "not spooky at all."[8] The weekly flight between Amsterdam and Leningrad launched in 1968 was just one sign of an increased interest in life behind the Iron Curtain, which attracted more tourists to the Soviet Union.

Biographical approach in the history of tourism

This chapter situates Pool's comments on the Soviet Union within her diverse positioning, and sheds a light on (local) power relations.[9] The focus on individual travellers makes not only "invisible" subjects visible, but also their individual and subjective perceptions, which can counter standardised tourist narratives.[10] By acknowledging differences among tourists, this biographical case study also shows that the personal is inherently political.

Far from being passive spectators, even small interactions between tourists and locals can constitute cultural production, as tourism co-creates liminal, in-between spaces that can contest their world and the other(s).[11] Although such individual travels might seem marginal, and thus insignificant, individual travels can expose both travellers and locals to different worldviews. In her 1968 autobiography, Pool explained her role as a temporary visitor in the American South. Locals often told their deepest personal secrets to her, perhaps because she "vanished to some place three or four thousand miles away" after a few weeks – back to London. Her in-between position as a "non-dangerous witness" meant that people confided in her. "Secrets are safe with me," was her explanation.[12] Clearly it is worthwhile to delve further into her short-stay holidays, and not just her longer trips.

Travel in the Soviet Union during the Cold War was symbolic and filled with significance. Especially for political tourists and (closet) homosexuals, it could be a transformative experience, a pilgrimage, and almost invigorating.[13] This chap-ter will explore Pool and Isenburg's encounters, looking for insights into tourist interactions with Soviet citizens.

Travellers and fellow travellers

From the mid-1920s to the 1960s Pool and Isenburg were fellow travellers: part of a broad group of Western left-wing intellectuals who shared a critical stance towards Western society, and who were broadly sympathetic towards socialism, and sometimes communism and the USSR too.[14] Pool had experienced the inter-war Popular Front movements, and especially during the war she leaned heavily towards communism.[15] Although Pool never became a party member, her experiences with the old left, in an anti-fascist resistance group during the war and as a Holocaust survivor, left her sympathetic to communist beliefs.

Still, Pool's interest in Russia and the Soviet Union had been marginal up to that point. She had translated some of Tchaikovsky's correspondence and learned some Russian phrases to entertain friends at cocktail parties. That makes it difficult to pinpoint what she "truly" felt about the Soviet Union around 1965. As Pool mostly operated in (African-)American circles, predominantly on the left, her behaviour was firmly rooted in a specific US context of the 1950s. According to historian James Smethurst, Cold War–era leftists adapted a pattern of locution, as they "generally assumed that their correspondence was being read at least occasionally by various intelligence agencies."[16] Leftist intellectuals like Pool self-censored their public and personal writings accordingly. With the political relationship between East and West changing frequently during Khrushchev's leadership (1956–64), political outspokenness remained an ongoing concern for many fellow travellers and leftists. And even when the immediate threat of McCarthyism was gone, many leftists remained cautious throughout the 1960s. Smethurst calls Pool's correspondence "fascinating documents of Cold War political circumspection," as "the correspondents cautiously come out of their political closets," mostly by mentioning the names (W. E. B. Du Bois, Shirley Graham, David Du Bois, Walter Lowenfels) that revealed their sympathies, while not revealing too much at the same time.[17] And in her public appearances, Pool often concealed her former socialist and communist sympathies. As a result she chose not to publish travel accounts from her experiences in the Soviet Union, even though she consistently published accounts of all her other journeys. This forced me to read her and Isenburg's few comments in their letters "against the grain" in order to detect subtle or hidden messages.[18]

This "other" reading makes it clear that the visit had a great impact. Pool wrote to a friend afterward that "The USSR . . . was a great experience and killed many prejudices."[19] One of those prejudices was about Siberia: "we almost passed out with the heat in Siberia," Pool wrote afterwards, "although one's association with Siberia is snow and ice."[20] Not even fellow travellers were immune to negative stereotypes about the Soviet Union. Elsewhere, the two women used positive remarks about the Soviet Union to affirm their political sympathies, while carefully mentioning public figures in order to maintain plausible deniability.[21]

Interesting conversations, fine scenery

From 24 May to mid-June 1965, Pool and Isenburg enjoyed a four-week holiday in the Soviet Union. They travelled around the northern half of the globe, departing from the US, via Japan across Asia towards London – an impressive 13,000 miles. The majority of the trip was spent in the Soviet Union, which they travelled "to the (almost) entire length and width," Pool wrote, "from Nakhodka (near Vladivostok) all the way through Siberia and Ural [*sic*], to Moscow, to the Crimes [*sic*], and Caucasus, [the] Republic of Georgia and onward to Leningrad and home by Russian boat from there to London." This journey included the legendary Trans-Siberian Railway between Irkutsk and Moscow. Pool described the train journey as "some of the most revealing 3½ days in my life, wonderful people

Figure 6.2 Pool (left) and Isenburg, summer of 1947

Source: © Rudi Wesselius

met, interesting conversations, fine scenery."[22] Pool, fluent in Dutch, German, and English, and with a working knowledge of French, Italian, and Norwegian, had memorised some Russian words and phrases so she could engage in small talk with locals. They communicated with other passengers by using words from various languages: "I made myself nicely misunderstood with [a] little Russian, their smatterings of English, French and German, and a dictionary."[23] If that did not help, they used hand gestures.

It is hard to see this rudimentary communication as an actual exchange with Soviet citizens. Yet, Pool experienced the train ride as an almost transformative experience: "The four days in that small railway community from Irkutsk to Moscow were an education in itself." Nevertheless, the limitations were obvious and frustrated the two women: "How many times I exclaimed, heavens to Betsy, how do I explain that to them!"[24] Pool's story suggests that there were mainly Russian travellers on the train who were friendly towards these tourists. In fact, the two women travelled the way Soviet citizens liked it best: in an educational and useful manner, travelling from one place to another,[25] which probably contributed to the kindness they encountered.

A bourgeois crime: lesbians in the Soviet Union

Their cautious attitudes with American correspondents was also practised in the flesh during their holiday in the Soviet Union, where they also had to be careful, but now on a completely different level as lesbian travellers. It is unlikely that Pool and Isenburg highlighted their sexuality during these encounters – not only were they unable to do so due to language barriers, but it was also wise to keep silent. Despite a lack of historical research on lesbian tourism, it is safe to say that Pool and Isenburg had to be extremely careful.[26]

Homosexuality was punishable in the Soviet Union from the 1930s onwards. By the 1960s it was seen as "shameful" or a "bourgeois crime," but most of all it was considered perverse.[27] The era of de-Stalinisation was actually one of the most conservative periods in Soviet society for homosexuals. The 1960s showed a sharp increase in annual convictions for sodomy; same-sex relations between women were known as *vialotekushaia shizofreniia* ("sluggishly manifesting schizophrenia").[28] Although not punishable by law, lesbians could be brought before "comrades' courts," where if convicted they could be sentenced to forced hospitalisation, psychiatric drugs, psychological therapy, or the loss of some civil rights.[29] As a result, lesbianism was surrounded by a "conspiracy of silence" until the 1980s, and lesbians remained largely invisible.

As tourists moved around in-between spaces they remained largely invisible as well, especially to the comrades' courts. Pool and Isenburg probably went along with the "silence" on the topic. It must have been a familiar situation. During their American travels, Pool often said that she was accompanied by her "best friend" – not an uncommon thing for lesbians to say on either side of the Iron Curtain. The attitude they adopted during their holiday was thus perhaps not so different from their ordinary lives in the UK. Even back in London, they often left their relationship unspecified, because Pool, according to a family member, was "a bit prude." Pool despised the Dutch word *vriendin* (girlfriend); it was too "piquant" to her taste.[30]

Despite her secrecy, Pool has been described as an advocate of lesbian and gay emancipation.[31] Some of the reasons the Soviet Union appealed to many left-wing Westerners was because of its idealised progressive stance toward homosexuals (largely untrue) and towards gender equality (partially true, for although Soviet women in general had better career options than their peers in the West, they experienced a double burden of work and housework).[32] This image of gender equality also attracted many (non-political) female travellers. From the 1920s to the 1940s, almost 40 per cent of tourists to the Soviet Union were women travelling independently.[33] Two "single" middle-aged women travelling together would not have been all that strange.

In this respect Pool and Isenburg were not so uncommon. However, the route they took, the distance they covered, and the means of transportation were more unusual. A month-long holiday was considerably longer than most Vernu-reizen trips (which usually lasted seven to twelve days).[34] Moreover, the return from Leningrad to London was made by boat instead of the usual train journey via Berlin: "[I completed] the cycle," Pool wrote to a friend, "by crossing over from the latter city [Leningrad] to London in a magnificent new Russian ship via Helsinki and Gothenburg in Sweden."[35] This made their holiday considerably more expensive, although it was not an insurmountable problem. After all, their sexual identity not only imposed limitations, it also offered opportunities. They were two working women without children, and thus had a considerable disposable income to travel "the other half of our magnificent world."

A country without racism: African Americans and the Soviet Union

Since the 1920s Pool had been an advocate of black emancipation. After the war she never hesitated to compare European anti-Semitism with American racism, often saying that racial segregation was a first step towards another Holocaust.[36] Pool was thus a vocal critic of Western racial attitudes, comparable to other fellow travellers and radical African-American intellectuals. Many of the latter regarded the Soviet Union as a promised land.[37] Some even described it as "the country without racism" where there was "no Jim Crow."[38]

Pool was certainly aware of the special place that the Soviet Union had in the minds of many African American intellectuals. In 1962 she wrote a critical article for a Soviet women's magazine on American racial segregation.[39] She also corresponded with many African Americans who had visited or would visit the Soviet Union, including the poet Langston Hughes (in 1932–33), the publisher Dudley Randall (in 1966), Vera C. Foster of the Tuskegee Institute, Alabama (in 1964), Charles "Chuck" Anderson (probably 1963), and the poet and artist Margaret Burroughs (in 1965 and 1966).[40] Some of them were fellow travellers, some outspoken communists. The famous scholar W. E. B. Du Bois and his wife, the playwright Shirley Graham, went to stay with Pool in London before continuing their trip to the Soviet Union and China in 1958.[41]

It is not unthinkable that Pool visited a number of well-connected Soviet citizens of African and African American descent, such as Lloyd Patterson, who lived in Komsomolsk-na-Amure in the Russian Far East. Perhaps she visited the Afro-Soviet historian Lily Golden (1934–2010) in Moscow, whom she knew personally.[42] Possibly they even met Margaret Burroughs (1915–2010), who was also travelling in the Soviet Union in 1965, the country where her husband had spent his childhood.[43] However, there are no records of such meetings. Contacts with American progressive radicals could be dangerous if the couple ever wanted to visit the US again. McCarthyism was a thing of the past by the mid-1960s, but since several of her close friends had been questioned by the McCarthy committee, the episode was still fresh in everyone's minds. The political tide could easily turn against progressive leftists again. The majority of their trip went unmentioned in their letters to friends. A short reference to Tashkent (in Uzbekistan), which was not on their route, shows that they kept a good deal of their experiences to themselves. Nevertheless, Pool wrote proudly to some friends and acquaintances that she had visited the USSR.[44] In anti-racist circles it was certainly no bad thing to have seen this "promised land" with her own eyes.

Museums of religion: Baha'is and Jews in the Soviet Union

Pool and Isenburg both had similar multilayered identities that usually remained invisible, especially as tourists passing through. Both women were secular Jews, but in the 1960s they started a spiritual quest. They eventually found religious fulfilment in the Baha'i faith, which they joined in May 1965 while in Huntsville, Alabama, shortly before travelling on to the USSR. The Baha'i faith, a monotheistic religion that stresses the spiritual unity of all humankind, originated in Persia in the nineteenth century.[45] Ever since the Baha'i faith spread in Russia in the 1880s, its supporters there had been severely persecuted.[46] In 1928 the Baha'i faith was seen as "anti-Soviet," and the government decided that it had an "espionage character" and therefore needed to be wiped out within the borders of the USSR.[47] The religion was forbidden in 1938.

The two new converts kept an eye open for other Baha'is, whom Pool regarded as her new "family." After her conversion she visited Baha'is all across the world, starting immediately in Japan, at the beginning of their trip. In Kyoto, Pool had a phone call with American–Hawaiian author and Baha'i follower Agnes Alexander (1875–1971) before heading into the city: "We spent most of our time in Kyoto visiting shrines and temples and people outside, and the evenings in the city," Pool wrote. "Fascinations [*sic*], enormously fascinating all of it."

Religion thus became an important focal point for their Russian trip. In the mid-1960s there were only two small isolated Baha'i communities left (in Tashkent and Fergana, both in Uzbekistan), about 200 Russian Baha'is in total.[48] However, Pool remained largely uncritical of their oppression; instead, she praised the tolerance of religion in the USSR:

Then the trek across Russia, four weeks on Russian soil. We generally found much more interest in religious matters than we anticipated. Even the fact that so many people come to see the 'museums' of religion, which are established in inactive churches, is significant. I found several fully attended orthodox churches, a very full mosque in Leningrad (a beauty too, true copy of the ruined one of Tashkent).[49]

Pool was clearly amazed that religious traditions had survived in a "disenchanted" Soviet Union. Was it the amazement of a Western traveller, or a praise from a fellow traveller? Her admiration for the tolerance of the Soviet government towards religious "dissidents" was unmistakable, though. The government not only tolerated the existence of churches as "museums" – a thing of the past – but also let people practise their religion. Her remarks amounted to support for the Soviet Union. The Western world frequently reported on the persecution of religion in the USSR.[50] This intolerance was indeed the case in the Stalinist period. Despite it, religious practices, faith, and discourses became more widespread even during the anti-religious campaign in the 1950s.[51] In fact, after Brezhnev came to power in 1964 both private and public religion was indeed tolerated, or at least ignored. It was not full acceptance of the kind Pool seemed to suggest, however.

In that same letter, Pool also referred to the position of the Jews, another persecuted minority. Pool noted:

mass attendance at an open air concert in a park in Sochi in the Caucasus for artists from Vilnius, Jews, singing Yiddish, and Hebrew, even songs to beautiful texts of Bialik, first in Hebrew, then in Yiddish, and from the laughter about jokes in folksongs, wedding tunes etc. one could gather how well understood it all was.[52]

Historians have looked askance at the non-critical stance of many fellow travellers towards the USSR: many praised Soviet society in every possible way in their travel reports, while being highly critical of Western regimes.[53] This unquestioning and naïve stance on the Soviet Union is also evident in Pool's reflections on her journey: "Much more freedom of thought than expected there, the people opener, more outgoing, warmer, welcoming beyond expectation."[54] This obviously raises questions. How could Pool ignore the discrimination, the persecution of both Baha'is and, more subtly, the Jews in the Soviet Union?

Upon closer inspection a different reading becomes possible. Her mention of Hebrew, a marginalised language,[55] the seemingly casual mention of Bialik, was not just a coincidence; it was Pool showing her political colours. The fact that she referred to the liberal Jewish poet Hayim Nahman Bialik (1873–1934) was especially telling. Bialik, the so-called "national poet" of the Zionist movement, condemned anti-Semitic violence and was forced to live for many years in exile in Germany and Palestine. Pool's suggestion that such dissident voices were accepted, and that modern Hebrew was spoken in the Caucasus, was a sign that she opposed the Western view of the USSR as one monolithic bloc. To that extent she debunked the myths about socialism, although very carefully, and invisibly to the untrained eye.

Concluding remarks

Pool and Isenburg interpreted the Soviet Union within their own frameworks as Western European, post-1956 (former) fellow travellers. Like other fellow travellers, Pool always sympathised with the underdog – in this case the Soviet Union itself – and dismissed negative images of the Soviet Union by only paying attention to the positive aspects of Soviet society. Western leftists of this type saw the USSR as a country that offered (greater) equality for Blacks, women, and other minorities. Paradoxically, the minority groups that Pool and Isenburg could relate to (lesbians, Baha'is, Jews) often remained invisible in Soviet society because they were persecuted.

Since tourists inhabit an in-between space, Pool and Isenburg often remained invisible as well. Communication consisted of a few words of Russian, other European languages, and some basic expressive gestures. The cultural exchange was thus extremely limited. Her background and the lack of communication made Pool very uncritical, perhaps even naïve, especially when compared to her opinion on Western societies.

Yet their remarks in their letters, read against the grain, reveal some subtle or hidden messages. When she mentioned that many of their prejudices about Siberia disappeared, and that the train ride was a transformative experience, they were cautiously positive remarks that enabled her to maintain plausible deniability. The mentioning of poet-in-exile Bialik was one of the most outspoken of her comments. A close reading includes reading between the lines, and gives valuable information about what could be implied by tourists to the Soviet Union.

Notes

1 Rosey Pool to AMSAC, 24 August 1965, Box 1, American Society of African Culture (AMSAC) Collection, Howard University, Moorland-Spingarn Research Center (MSRC).
2 A religion tolerant of all other world religions, emphasising the unity and equality of all people.
3 Rosey Pool to Langston Hughes, 26 July 1953, Box 130, Folder 2430, Langston Hughes Papers, Beinecke Rare Book & Manuscript Library, Yale University.
4 Rosey Pool to AMSAC, 12 November 1963, Box 1, AMSAC Collection, MRSC.
5 Rosey Pool to Margaret 'Mother' McCall, 1 September 1964, Folder 99, Rosey E. Pool Papers, Howard University MSRC.
6 "Rapportje bezoek Moskou," 23–30–12/65, inv.no. 89: Correspondence Vereniging Nederland – USSR (VERNU), 1966, Archive Wim Hulst, International Institute of Social History (IISH).
7 "Duizenden Nederlanders met vakantie naar Sovjet-Unie: Reizen achter IJzeren Gordijn zijn 'in,'" *Nieuwsblad van het Noorden*, March 26, 1965, 39.
8 "Niets griezeligs aan toerisme in de Sowjet-Unie: Russen zien ons graag komen," *Het Parool*, June 26, 1965, 4, inv.no. 86, Archive Wim Hulst, IISH.

9 Kathy Davis, "Intersectionality as Buzzword: A Sociology of Science Perspective on What Makes a Feminist Theory Successful," *Feminist Theory* 9, no. 1 (2008): 70.

10 Cord Pagenstecher, "Zwischen Tourismuswerbung und Autobiographie: Erzählstrukturen in Urlaubsalben," in *Gebuchte Gefühle: Tourismus zwischen Verortung und Entgrenzung*, ed. Hasso Spode and Irene Ziehe (Munich: Profil, 2005), 82–91.

11 Homi K. Bhabha, *The Location of Culture* (Abingdon: Routledge, 2012 [1994]), 5; Babs Boter, "Heavenly Sensations and Communal Celebrations: Experiences of Liminality in Transatlantic Journeys," in *Tales of Transit: Narrative Migrant Spaces in Atlantic Perspective, 1850–1950*, ed. Michael Boyden, Hans Krabbendam and Liselotte Vandenbussche (Amsterdam: Amsterdam University Press, 2013), 180.

12 Rosey Pool, *As Waves of One Sea* [English translation of *Lachen om niet te huilen*], ch. "Creativity," 3, SxMs19/11/3/18, Rosey Pool Collection, University of Sussex, Special Collections at The Keep.

13 Aaron Lecklider, "TWO Witch-hunts: On (Not) Seeing Red in LGBT History," *American Communist History* 14, no. 3 (2016): 246; Paul Hollander, *Political Pilgrims: Travels of Western Intellectuals to the Soviet Union, China, and Cuba, 1928–1978* (New York: Oxford University Press, 1981), 38; Garth L. Lean, "Transformative Travel: A Mobilities Perspective," *Tourist Studies* 12, no. 2 (2012): 151–72.

14 David Caute, *The Fellow-Travellers: Intellectual Friends of Communism* (New Haven: Yale University Press, 1988 [1973]); Max van Weezel and Anet Bleich, *Ga dan zelf naar Siberië! Linkse intellektuelen en de koude oorlog* (Amsterdam: SUA, 1978).

15 Pool's resistance group involved many party communists and sympathizers. See Milo Anstadt, *Kruis of munt: Autobiografie 1920–1945* (Amsterdam: Contact, 2000), 343; Ben Braber, "Passage naar vrijheid: De groep-Van Dien: Duitse joden in Nederlandse illegaliteit" (MA diss., University of Amsterdam, 1986), 82. After the war, Pool also held recitals of African American poetry in anti-fascist and communist clubs in Amsterdam, like the newspaper *De Vrije Katheder*. See Peter van Steen, "Rosey Pool draagt negergedichten voor," *De waarheid*, October 15, 1945, 3; Fenna van den Burg, "De Vrije Katheder: Een platform van communisten en niet-communisten, 1945–1950" (PhD diss., University of Groningen, 1983).

16 James Smethurst, "'Don't Say Goodbye to the Porkpie Hat': Langston Hughes, the Left, and the Black Arts Movement," *Callaloo* 25, no. 4 (2002): 1226; James Smethurst, "The Black Arts Movement and Historically Black Colleges and Universities," in *New Thoughts on the Black Arts Movement*, ed. Lisa Gail Collins and Margo Natalie Crawford (New Brunswick: Rutgers University Press, 2006), 90f13.

17 Smethurst, "'Don't Say Goodbye to the Porkpie Hat,'" 122f13.

18 Stuart Hall, "Encoding, Decoding," in *Culture, Media, Language: Working Papers in Cultural Studies, 1972–1979*, ed. Centre for Contemporary Cultural Studies (London: Routledge 1980 [1973]), 128–38.

19 Rosey Pool to Tom Harris, 15 December 1965, Folder 60, Rosey Pool Papers, MSRC.

20 Anneke Buys, "The Marvellous Gift of Friendship: A Biography of Rosey E. Pool, 1905–1971" (Unpublished manuscript, Apeldoorn, 1986, n.p.), Report Rosey Pool on Guatemala, 27 April 1967.

21 Smethurst, "Don't Say Goodbye to the Porkpie Hat," 1226.

22 Rosey Pool to Ibrahim Ibn Ismail, 2 September 1965, folder 75, Rosey Pool Papers, MSCR.

23 Buys, 'The Marvellous Gift of Friendship', n.p. Rosey Pool to Florence and Stan Bagley, *c*.1965.

24 Ibid.

25 Diane P. Koenker, *Club Red: Vacation Travel and the Soviet Dream* (Ithaca: Cornell University Press, 2013), 210–11.

26 Research on lesbian tourists focuses mostly on contemporary examples. See J. K. Puar, "Circuits of Queer Mobility: Tourism, Travel, and Globalization," *GLQ: A Journal of Lesbian and Gay Studies* 8, no. 1 (2002): 101–37; Anette Therkelsen, Bodil Stilling

Blichfeldt, Jane Chor, and Nina Ballegaard, "'I Am Very Straight in My Gay Life': Approaching an Understanding of Lesbian Tourists' Identity Construction," *Journal of Vacation Marketing* 19, no. 4 (2013): 317–27.

27 Francesca Stella, "Lesbian Lives and Real Existing Socialism in Late Soviet Russia," in *Queer Presences and Absences*, ed. Yvette Taylor and Michelle Addison (Basingstoke: Palgrave Macmillan, 2013), 51; David Tuller, *Cracks in the Iron Closet: Travels to Gay and Lesbian Russia* (Chicago: University of Chicago Press, 1997), 5.

28 Dan Healey, *Homosexual Desire in Revolutionary Russia: The Regulation of Sexual and Gender Dissent* (Chicago: University of Chicago Press, 2001), 246. The unique Soviet term for lesbianism is perhaps comparable with the western 'borderline personality' disorder, see Laurie Essig, *Queer in Russia: A Story of Sex, Self, and the Other* (Durham: Duke University Press, 2012), 28.

29 Masha Gessen, *The Rights of Lesbians and Gay Men in the Russian Federation: An International Gay and Lesbian Human Rights Commission Report* (San Francisco: IGLHRC, 1994), 17–18.

30 Rosey Pool to Alice von Eugen [Querido publisher], n.d. [*c*.1953], 18–2 QUE Correspondence Pool, R.E., Museum of Literature, The Hague. For the translation see Catherine Drinker Bowen and Barbara von Meck, *Dierbare vriendin: De roman van Peter Tsjaikowsky en Nadesjda von Meck* (Amsterdam: NVEM Querido's Uitgeverij, 1953).

31 Rudi Wesselius, email to author, 15 December 2015. Pool wrote articles for the Dutch homosexual organization COC, *Vriendschap*, and she had many people in her network who (secretly) identified as homosexual or at least bisexual, including Albert Mol, Langston Hughes, Owen Dodson, Countee Cullen, and Harold Jackman.

32 For the situation for homosexuals in the USSR, see Brian James Baer, "Queer in Russia: Othering the Other of the West," in *Queer in Europe: Contemporary Case Studies*, ed. Lisa Downing and Robert Gillett (Burlington: Ashgate, 2011), 177. For the position of women in the Soviet Union, see Barbara Evans Clements, *A History of Women in Russia: From Earliest Times to the Present* (Bloomington: Indiana University Press, 2012), 261; Barbara Engel, "Engendering Russia's History: Women in Post-Emancipation Russia and the Soviet Union," *Slavic Review* 51, no. 2 (1992): 319.

33 Sheila Fitzpatrick and Carol Rasmussen, eds., *Political Tourists: Travellers from Australia to the Soviet Union in the 1920s–1940s* (Melbourne: Melbourne University Press, 2008), 25–26; Emily Lygo, "Promoting Soviet Culture in Britain: The History of the Society for Cultural Relations Between the Peoples of the British Commonwealth and the USSR, 1924–45," *Modern Language Review* 108, no. 2 (2013): 579.

34 Correspondence Vereniging Nederland-USSR (VERNU), 1963, inv.no. 86, Archive Wim Hulst, IISH.

35 Rosey Pool to Samuel Boyea, 2 September 1965, folder 17, Rosey Pool Papers, MSRC.

36 William A. Fowlkes, "Author Pool Notes Similarity Of Nazi Oppression, Segregation," *Atlanta Daily World*, February 12, 1960, 1, SxMs19/14/1/4 Scrapbook 1959–1960, Rosey Pool Collection, University of Sussex.

37 Kate A. Baldwin, *Beyond the Color Line and the Iron Curtain: Reading Encounters Between Black and Red, 1922–1963* (Durham: Duke University Press, 2002); Joy Gleason Carew, *Blacks, Reds, and Russians: Sojourners in Search of the Soviet Promise* (New Brunswick: Rutgers University Press, 2008); Maxim Matusevich, "Journeys of Hope: African Diaspora and the Soviet Society," *African Diaspora* 1, no. 1 (2008): 53–85; Athan Andreas Biss, "Unexpected Frontiers of Black Internationalism: African Americans in Soviet Central Asia, 1930–1976," *Central Asian Affairs* 2, no. 2 (2015): 189–206.

38 Lily Golden, *My Long Journey Home* (Chicago: Third World, 2002), 20; Langston Hughes, "Negroes in Moscow: In a Land Where There is No Jim Crow," *International Literature*, no. 4 (1933): 78–81.

39 Rosey E. Pool, "In the Land of 'Whites' and 'Coloured'," *Soviet Woman*, no. 2 (1962): 28–29.

40 Pool's correspondence with these individuals can be found at the Rosey E. Pool Papers (situated at Howard University) and the Rosey Pool Collection (University of Sussex).

41 15.5f. W.E.B. Du Bois scrapbook: trip to Holland, 1958, Folder 15.4, MC 476, Papers of Shirley Graham Du Bois, 1865–1998, Arthur and Elizabeth Schlesinger Library on the History of Women in America, Radcliffe Institute for Advanced Study, Harvard University.

42 Lonneke Geerlings, "De Sovjet-Unie, 'the Country Without Racism': Een Portret van de Afro-Russische Historica Lily Golden (1934–2010)," *Historica: Tijdschrift voor gendergeschiedenis* 39, no. 3 (2016): 12–17.

43 Margaret Taylor Burroughs, "My Dream Has Come True," *Soviet Woman*, no. 2 (1965): 12; Mary Jane Dickerson, "Margaret T. G. Burroughs," in *Afro-American Poets since 1955*, ed. Trudier Harris and Thadious M Davis (Detroit: Gale Research, 1985), 52; Margaret T. G. Burroughs, *Life with Margaret: The Official Autobiography* (Chicago: In Time, 2003).

44 Pool writes about her trip to Jessie Collins, Samuel Boyea, Arnold Grimes, Tom Harris, Jean Blackwell Hutson, Ibrahim Ibn Ismail, and others.

45 Moojan Momen and Peter Smith, "The Baha'i Faith 1957–1988: A Survey of Contemporary Developments," *Religion* 19, no. 1 (1989): 71.

46 Walter Kolarz, *Religion in the Soviet Union* (London: Macmillan, 1962), 470–73. In general, religion was frowned upon by communist ideology, but the Baha'i faith especially aroused suspicion because, contradictorily, of its broadmindedness, its tolerance, its international outlook, and its emphasis on gender equality and female emancipation. This infuriated Soviet officials because "All this contradict[ed] the communist thesis about the backwardness of all religions," the British scholar Walter Kolarz duly remarked in 1962.

47 Felix Corley, *Religion in the Soviet Union: An Archival Reader* (Basingstoke: Macmillan, 1996), 60.

48 "Statement on the History of the Bahá'í Faith in Soviet Union," official website of the Baha'is of Kiev, August 2007, https://web.archive.org/web/20100715093037/ http://bahai.kiev.ua/history9.html.

49 Buys, *The Marvellous Gift of Friendship*, n.p.; Rosey Pool to Florence and Stan Bagley, *c*.1965.

50 Mariam Dobson, "The Social Scientist Meets the 'Believer': Discussions of God, the Afterlife, and Communism in the Mid-1960s," *Slavic Review* 74, no. 1 (Spring 2015): 83.

51 Stéphane A. Dudoignon and Christian Noack, eds., *Allah's Kolkhozes: Migration, De-Stalinisation, Privatisation, and the New Muslim Congregations in the Soviet Realm (1950s–2000s)* (Berlin: KS, Klaus Schwarz, 2014), 12; Ulrike Huhn, "Die Wiedergeburt der Ethnologie aus dem Geist des Atheismus: Zur Erforschung des 'zeitgenössischen Sektierertums' im Rahmen von Chruščevs antireligiöser Kampagne," *Jahrbucher fur Geschichte Osteuropas* 64 no. 2 (2016): 260–98.

52 Buys, *The Marvellous Gift of Friendship*, n.p.; Rosey Pool to Florence and Stan Bagley, *c*.1965.

53 Aart Aarsbergen, *Verre paradijzen: Linkse intellectuelen op excursie naar de Sovjet-Unie, Cuba en China* (Utrecht: HES, 1988), 13.

54 Buys, *The Marvellous Gift of Friendship*, n.p.; Rosey Pool to Florence and Stan Bagley, *c*.1965.

55 Zvi Gitelman, *A Century of Ambivalence: The Jews of Russia and the Soviet Union, 1881 to the Present* (New York: Schocken, 1988), 247.

References

Aarsbergen, Aart. *Verre paradijzen: Linkse intellectuelen op excursie naar de Sovjet-Unie, Cuba en China*. Utrecht: HES, 1988.

Anstadt, Milo. *Kruis of munt: Autobiografie 1920–1945*. Amsterdam: Contact, 2000.

Baer, Brian James. "Queer in Russia: Othering the Other of the West." In *Queer in Europe: Contemporary Case Studies*, edited by Lisa Downing and Robert Gillet, 172–88. Burlington: Ashgate, 2011.

Baldwin, Kate A. *Beyond the Color Line and the Iron Curtain: Reading Encounters Between Black and Red, 1922–1963*. Durham: Duke University Press, 2002.

Bhabha, Homi K. *The Location of Culture*. Abingdon: Routledge, 2012 [1994].

Biss, Athan Andreas. "Unexpected Frontiers of Black Internationalism: African Americans in Soviet Central Asia, 1930–1976." *Central Asian Affairs* 2, no. 2 (2015): 189–206.

Boter, Babs. "Heavenly Sensations and Communal Celebrations: Experiences of Liminality in Transatlantic Journeys." In *Tales of Transit. Narrative Migrant Spaces in Atlantic Perspective, 1850–1950*, edited by Michael Boyden, Hans Krabbendam, and Liselotte Vandenbussche, 179–95. Amsterdam: Amsterdam University Press, 2013.

Bowen, Catherine Drinker, and Barbara von Meck. *Dierbare vriendin: De roman van Peter Tsjaikowsky en Nadesjda von Meck*. Translated by Rosey E. Pool. Amsterdam: NVEM. Querido's Uitgeverij, 1953.

Braber, Ben. "Passage naar vrijheid: De groep-Van Dien: Duitse joden in Nederlandse illegaliteit." MA diss., University of Amsterdam, 1986.

Burg, Fenna van den. "De Vrije Katheder: Een platform van communisten en niet-communisten, 1945–1950." PhD diss., University of Groningen, 1983.

Burroughs, Margaret T. "My Dream Has Come True." *Soviet Woman*, no. 2 (1965): 12.

———. *Life with Margaret: The Official Autobiography*. Chicago: In Time & Media, 2003.

Buys, Anneke. "The Marvellous Gift of Friendship: A Biography of Rosey E. Pool, 1905–1971." Unpublished., Apeldoorn, 1986.

Carew, Joy Gleason. *Blacks, Reds, and Russians: Sojourners in Search of the Soviet Promise*. New Brunswick: Rutgers University Press, 2008.

Caute, David. *The Fellow-Travellers: Intellectual Friends of Communism*. New Haven: Yale University Press, 1988 [1973].

Clements, Barbara Evans. *A History of Women in Russia: From Earliest Times to the Present*. Bloomington: Indiana University Press, 2012.

Corley, Felix. *Religion in the Soviet Union: An Archival Reader*. Basingstoke: MacMillan Press, 1996.

Davis, Kathy. "Intersectionality as Buzzword: A Sociology of Science Perspective on What Makes a Feminist Theory Successful." *Feminist Theory* 9, no. 1 (2008): 67–85.

Dickerson, Mary Jane. "Margaret T. G. Burroughs." In *Afro-American Poets Since 1955*, edited by Trudier Harris and Thadious M. Davis, 51–54. Detroit: Gale Research, 1985.

Dobson, Mariam. "The Social Scientist Meets the 'Believer': Discussions of God, the Afterlife, and Communism in the Mid – 1960s." *Slavic Review* 74, no. 1 (2015): 79–103.

Dudoignon, Stéphane A., and Christian Noack, eds. *Allah's Kolkhozes: Migration, De-Stalinisation, Privatisation, and the New Muslim Congregations in the Soviet Realm (1950s–2000s)*. Berlin: KS, Klaus Schwarz, 2014.

Engel, Barbara. "Engendering Russia's History: Women in Post-Emancipation Russia and the Soviet Union." *Slavic Review* 51, no. 2 (1992): 309–21.

Essig, Laurie. *Queer in Russia: A Story of Sex, Self, and the Other*. Durham: Duke University Press, 2012.

Fitzpatrick, Sheila, and Carol Rasmussen, eds. *Political Tourists: Travellers from Australia to the Soviet Union in the 1920s–1940s*. Melbourne: Melbourne University Press, 2008.

Geerlings, Lonneke. "De Sovjet-Unie, 'the Country Without Racism': Een Portret van de Afro-Russische Historica Lily Golden (1934–2010)." *Historica: Tijdschrift Voor Gendergeschiedenis* 39, no. 3 (2016): 12–17.

Gessen, Masha. *The Rights of Lesbians and Gay Men in the Russian Federation: An International Gay and Lesbian Human Rights Commission report*. San Francisco: IGLHRC, 1994.

Gitelman, Zvi. *A Century of Ambivalence: The Jews of Russia and the Soviet Union, 1881 to the Present*. New York: Schocken, 1988.

Golden, Lily. *My Long Journey Home*. Chicago: Third World, 2002.

Gorsuch, Anne E. "'There's No Place Like Home': Soviet Tourism in Late Stalinism." *Slavic Review* 62, no. 4 (2003): 760–85.

———. *All This Is Your World: Soviet Tourism at Home and Abroad After Stalin*. Oxford: Oxford University Press, 2013.

Hall, Stuart. "Encoding, Decoding." In *Culture, Media, Language: Working Papers in Cultural Studies, 1972–1979*, edited by the Centre for Contemporary Cultural Studies, 128–38. London: Routledge, 1980 [1973].

Healey, Dan. *Homosexual Desire in Revolutionary Russia: The Regulation of Sexual and Gender Dissent*. Chicago: University of Chicago Press, 2001.

Hollander, Paul. *Political Pilgrims: Travels of Western Intellectuals to the Soviet Union, China, and Cuba, 1928–1978*. New York: Oxford University Press, 1981.

Hughes, Langston. "Negroes in Moscow: In a Land Where There Is No Jim Crow." *International Literature* 4 (1933): 78–81.

Huhn, Ulrike. "Die Wiedergeburt der Ethnologie aus dem Geist des Atheismus: Zur Erforschung des 'zeitgenössischen Sektierertums' im Rahmen von Chruščevs antireligiöser Kampagne." *Jahrbucher für Geschichte Osteuropas* 64, no. 2 (2016): 260–98.

Koenker, Diane P. *Club Red: Vacation Travel and the Soviet Dream*. Ithaca: Cornell University Press, 2013.

Kolarz, Walter. *Religion in the Soviet Union*. London: Macmillan, 1962.

Lean, Garth L. "Transformative Travel: A Mobilities Perspective." *Tourist Studies* 12, no. 2 (2012): 151–72.

Lecklider, Aaron. "TWO Witch-hunts: On (Not) Seeing Red in LGBT History." *American Communist History* 14, no. 3 (2016): 241–47.

Lygo, Emily. "Promoting Soviet Culture in Britain: The History of the Society for Cultural Relations Between the Peoples of the British Commonwealth and the USSR, 1924–45." *Modern Language Review* 108, no. 2 (2013): 571–96.

Matusevich, Maxim. "Journeys of Hope: African Diaspora and the Soviet Society." *African Diaspora* 1, no. 1 (2008): 53–85.

Momen, Moojan, and Peter Smith. "The Baha'i Faith 1957–1988: A Survey of Contemporary Developments." *Religion* 19, no. 1 (1989): 63–91.

Official Web Site of the Baha'i community in Kiev, Ukraine, "Statement on the History of the Bahá'í Faith in Soviet Union." August 2007. Accessed July 15, 2010 and April 16, 2019 https://web.archive.org/web/20100715093037/http://bahai.kiev.ua/history9.html. Archived from the original (https://bahai.kiev.ua/ru/history9.html).

Pagenstecher, Cord. "Zwischen Tourismuswerbung und Autobiographie. Erzählstrukturen in Urlaubsalben." In *Gebuchte Gefühle: Tourismus zwischen Verortung und Entgrenzung*, edited by Hasso Spode and Irene Ziehe, 82–91. Munich: Profil, 2005.

Pattle, Sheila Hellen. "'Tourism for Everyone': Domestic Tourism in the USSR During Late Socialism, 1950s–1980s." MA diss., Durham University, 2015.

Pool, Rosey E. "In the Land of 'Whites' and 'Coloured.'" *Soviet Woman*, no. 2 (1962): 28–29.

Puar, J. K. "Circuits of Queer Mobility: Tourism, Travel, and Globalization." *GLQ: A Journal of Lesbian & Gay Studies* 8, no. 1 (2002): 101–37.

Smethurst, James. "'Don't Say Goodbye to the Porkpie Hat': Langston Hughes, the Left, and the Black Arts Movement." *Callaloo* 25, no. 4 (2002): 1224–37.

————. "The Black Arts Movement and Historically Black Colleges and Universities." In *New Thoughts on the Black Arts Movement*, edited by Lisa Gail Collins and Margo Natalie Crawford, 75–91. New Brunswick: Rutgers University Press, 2006.

Stella, Francesca. "Lesbian Lives and Real Existing Socialism in Late Soviet Russia." In *Queer Presences and Absences*, edited by Yvette Taylor and Michelle Addison, 50–68. Basingstoke: Palgrave Macmillan, 2013.

Therkelsen, Anette, Bodil Stilling Blichfeldt, Jane Chor, and Nina Ballegaard. " 'I Am Very Straight in My Gay Life': Approaching an Understanding of Lesbian Tourists' Identity Construction." *Journal of Vacation Marketing* 19, no. 4 (2013): 317–27.

Tuller, David. *Cracks in the Iron Closet: Travels to Gay and Lesbian Russia*. Chicago: University of Chicago Press, 1997.

Weezel, Max van, and Anet Bleich. *Ga dan zelf naar Siberië! Linkse intellektuelen en de koude oorlog*. Amsterdam: SUA, 1978.

7 The Stalinist utopia of the Adriatic

Swedish tourists in communist Albania

Francesco Zavatti

> We did a telephone interview with the first group of travellers in Albania. However, the telephone just rang. Albania only has a telephone connection with the outer world at 10–10.45 and 14–15. Eventually we got in touch with our group leader there. . . . The water was 22 degrees, the air was 27 in the shadow . . . 'good' . . . 'shocking prices' . . . 'sun' . . . 'the Albanians.' The words bounced and crackled on the telephone. . . . You too can travel there now. Planes leave every Friday all summer long. The trips are cheap, because Albania is cheap and undiscovered.[1]

With these words, the Danish tourist bureau Danresor advertised to the Swedish public in May 1969 that a new tourist spot had just appeared on the map of Europe: Albania, an undiscovered tourist destination with pleasant summer weather. Such an isolated country, the advert suggested, was also far away from the inconveniences of mass tourism.

The advert was a deliberate simplification of the reality of such holidays. First, the official name of the country was absent: the state referred to as "Albania" was actually the People's Republic of Albania. Second, it did not mention that the organiser, Albturist, was the official travel agency of the People's Republic of Albania.[2] Both were deliberate attempts to downplay the socialist identity of Albania. The isolation had little to do with its geographical location, squeezed between Yugoslavia, Greece, and the Adriatic Sea. Instead, it owed everything to the Albanian communist leader Enver Hoxha's decision to cut off the "outer world." In the 1960s, Albania's only international alliance at a global level was with the anti-revisionist People's Republic of China.[3]

In the 1960s, Maoist China had become a new point of reference for the Western anti-revisionist Marxists-Leninists, who longed for liberation from the imperialism of both the Soviet Union and the US. Between 1967 and 1970, these Western radicals founded parties and friendship associations loyal to China. Because of its alliance with China, Albania had also become an international talking point – a close one, too, on the shores of the Adriatic Sea. The same radical groups founded several pro-Albanian parties and friendship associations during this period. In contrast to Mao's pragmatic international politics, Hoxha considered the Western European anti-revisionist parties an asset for Albanian foreign politics, and set about actively influencing them.[4] In this, the Albanian regime could count on the support of its Western sympathisers for institutionalising propaganda in the shape of "friendship visits" – organised tourist trips that were intended to create durable trans-systemic relations across the "semipermeable membrane" of the Iron Curtain.[5]

The dynamics of these trans-systemic propaganda initiatives raise questions about the expectations generated by the tourist adverts and the actual travel experiences. How was tourist travel in Albania sold, and what feedback did these initiatives receive from tourists? The aim of this chapter is thus twofold. First, it investigates the international propaganda initiatives of the Albanian communist regime by focusing on the tourist activity of one foreign "friendship" association; namely, the tourist packages organised, promoted, and implemented by the Swedish–Albanian Association (Svensk–Albanska Föreningen – hereafter the Association). These will be analysed with the support of travel adverts and sources from the Association's archive. Second, the chapter scrutinises the perception of Albania by Swedish participants in such friendship tours. Travel adverts and tourist feedback are the main sources for analysing how messages were propagated and how the audience reacted. By analysing the efforts of a relatively small and unknown friendship association in advertising holiday travels, it is possible to shed light on how trans-systemic propaganda initiatives worked and what results they achieved. The chapter argues that the tours were organised jointly, with the aim of establishing the international legitimacy of an otherwise politically isolated country. In advertising the package tours, the Association conveyed a dual image of Albania as a tourist paradise and as a successful revolutionary experiment. The attempt was not always successful once Swedish tourists could see Albania with their own eyes.

Organising friendship tourism

Members of KFML (Kommunistiska Förbundet Marxist-Leninisterna), a pro-Chinese splinter party of the Left Party–The Communists (Vänsterpartiet kommunisterna), established the Association in Stockholm in 1970.[6] One of the Association's main activities was the organisation of tourist travel, open to all Swedish citizens. Despite the Swedish radical left's political marginality and the irrelevance of Enverism as a political movement, the tourist activities of the Association are worth analysing for the numbers of tourists attracted every year to Albania, since they ranged from a few dozen to several hundred.[7]

Historians have pointed out that the Sweden–GDR Association (discussed by Birgitta Almgren) and the Swedish–Chinese Association (Anne Hedén) received financial contributions and propaganda materials from the socialist countries, which therefore could have influenced the associations' main activities and their cultural diplomacy in Sweden. The main tactic used by those two associations was to present themselves not as political movements, but as civil society organisations, open to whoever was interested in the language and culture of the "friendship country."[8] The associations translated, published, and distributed texts prepared in the friendship country, from literature to political propaganda; they organised study circles and cultural exhibitions on the history, politics, and culture of the country: organised travel, when it started, supplemented these activities, and was ostensibly open to anyone interested. The Swedish–Albanian Association did not differ from this model.

In 1969, one group of architecture students and one group of art students from the Royal Swedish Academy of Fine Arts travelled to the People's Republic of Albania.[9] After their journey, some of them advertised the forthcoming establishment of a Swedish–Albanian Association.[10] The sources do not tell us if anyone responded to the invitation. However, in 1970, those early visitors to communist Albania edited the book *Albanien* ('Albania'), published by the newly founded socialist press, Ordfront.

Some of them succeeded in creating enduring relations with the Albanian regime. The Albanian Committee for Cultural and Friendship Relations with Foreign Countries, an Albanian state organ, suggested the creation of a friendship association to some of those early "friends of Albania" (Albanienvänner), who reacted positively in March 1970.[11] At that stage, the Association could count on the sympathies of pro-Albanian communists and Maoists more generally, but it also attracted the support of some of the politically aligned artists and architects who had visited the country the previous year. Per-Olof Ultvedt, an internationally renowned artist and professor at the Royal Academy, who had led his students to Albania on the trip of 1969, accepted the role of secretary-general of the Association.

Enthusiasm for Albania in the radical left milieu encouraged about two hundred people to book holidays organised by *Gnistan*, the KFML periodical. The Association supported *Gnistan*'s trips by setting up study circles on the Albanian economy, politics, literature, and language for the tourists, so that they would arrive prepared for their daily study visits.[12] For the Association, inviting people to see Albania with their own eyes and to experience it at first hand was the most powerful means of spreading knowledge about it in Sweden. The destiny of the Association, like that of the image of the People's Republic of Albania, hinged on the satisfaction of those Swedish tourists.

In April 1970, the Association took responsibility for organising a group of twenty to visit in August. This was a trial, with a view of taking over responsibility for all the Swedish package holidays to Albania. The Association chose to work with Spies Rejser, a Danish travel agency based in Copenhagen. Initially, it had wanted Albturist, as in 1969, but interestingly, Albturist had suggested that the Association would do better to deal with a larger travel agency.[13] This "suggestion" probably reflected the interests of the Albanian regime (to whom Albturist belonged), which sought cooperation with bigger commercial travel agencies.[14] Hence the Association chose the comparatively cheap deals offered by Spies Rejser.[15]

The adverts targeted both leisure tourists and political pilgrims by focusing on the leisure opportunities and the political peculiarities of the Albanian regime. The Association travel brochure for 1970 included "visits to factories, farms, schools, nurseries, news bureaus, publishing houses, other activities at a house of culture . . . genuine popular music shows or meeting someone who could speak on areas such as the economy, culture, education, and to whom we can put questions."[16] The board of the Association was conscious that the very different objectives of political pilgrimages and leisure tourism had somehow to be reconciled. Since the tourists wanted (as always) a relaxing beach holiday, an important task for Association members, as stated in an internal document in the Association's archive, was to "relate with those who want [only] to bathe and to rest."[17] The tactic was to turn politically indifferent tourists into people interested in Albania and the Albanians, with the ultimate goal of persuading them to join the Association.

After the trial trip in 1970, the Association was ready to take over responsibility for all trips to Albania under the aegis of *Gnistan*, and thus of turning tourists into friends of Albania.[18] The Association had the monopoly on Swedish package tours to Albania, and this allowed it to take 200 Swedish tourists to Albania in 1971. From the available data, it seems that the Association's activism paid off in terms of membership. In 1972, the Association grew from 385 to 625. In the same year, the first Albanian embassy in a Nordic country opened in Stockholm. Consequently, the Association invited representatives of the Finnish, Danish, and Norwegian friendship associations to Stockholm in order to meet the Albanian ambassador, Sami Baholli.[19] During that meeting, the Association took on the coordination of tourist travel for all the Albanian friendship associations in the Nordic countries.[20]

The presence of the Albanian embassy in Stockholm and the collaboration with the other Nordic associations coincided with the search for a new business partner. From 1973, the Association organised its tours with Reso, the Swedish labour movement's travel agency. Reso offered accommodation with full board in two different hotels in Durrës: the Association could thus diversify a little. At the same time, however, participation was now limited to those who had been members for at least a year. Officially, the restriction was intended to strengthen the study-trip character of the visits. Yet it is also possible that it had been prompted by a demonstration by one of the 1971 tourists: Wilgot Fritzon, a theology student, had arrived in Albania with forty Bibles in Albanian and some pamphlets with the schedule for the Christian radio Monte Carlo, which he handed out around Tirana. Once discovered, he admitted responsibility and was immediately expelled from the country.[21] The change in policy may also have been intended to improve members' chances of visiting the country, as due to the Association's growth, many of them had not yet had the opportunity; however, non-members were accepted for tours in subsequent years, even though the number of members continued to grow, reaching 896 in 1976.[22]

In 1978, the breaking of the Sino-Albanian alliance had consequences for the Swedish left. The Maoists left the Association, cutting its membership to 350.[23] Travel to Albania continued nonetheless. In 1979, 90 tourists visited Albania with the Association, while in 1980 there were 150.[24] The devaluation of the Swedish krona in 1981–82 sharply increased the cost of foreign travel.[25] To reduce its prices, the Association planned to travel by train and bus, and shared visits with the other Nordic associations. All this led to a sudden fall in tourist numbers, and only 24 people travelled to Albania in 1982.[26] From 1984 until the late 1980s, package tours continued in collaboration with the other Nordic associations. Despite the organisational adjustments, they attracted only 40 or so a year.[27] After the death of Enver Hoxha in 1984, the new leader Ramiz Alia decided in 1986 to improve relations with Western countries. Consequently, in 1987 the Association and its business partners lost their monopoly on travel to Albania. Another travel agency, Fritidsresor, immediately became the main competitor, and succeeded in attracting 700 tourists in the summer of 1987.[28] Meanwhile, the number of tourists travelling with the Association fell even further in subsequent years.

Tourist paradise, Stalinist utopia

From the start the Association advertised Albania both as a tourist paradise and as the realisation of a Stalinist utopia. In November 1970, the Association launched a publicity campaign in Stockholm in cooperation with KFML and the socialist organisation Clarté. Its advertising paid off. With the help of public events on Albanian culture, art exhibitions, the distribution of the Albanian bimonthly *New Albania*, and publication of the first travel report, *Albansk Utmaning* (*Albania Defiant*) by its famous members Gun Kessle and Jan Myrdal, the Association succeeded in recruiting 118 new members. The adverts promoted the revolutionary struggles of the Albanian comrades and presented the country as a "China in miniature."[29] It promised that travellers would meet Albanians going about their everyday lives, and that they would take part, for one day, in voluntary work, building a railway.

When the Association took over from Clarté in 1971, it attempted to sell the visits to Albania both as travels into an "interesting country from which even the Swedish people can learn" and as "a restful relaxation from everyday life."[30] The visits included excursions to natural wonders and cultural heritage sites that the regime considered worthy of international promotion: the coastal city of Vlora, the archaeological ruins of Butrint, and the historical city of Girokaster. In addition, as extras the tourists could visit the capital Tirana with its museums and shopping centres, the old capital Kruja with its castle, the ancient towns of Skhodra and Berat, the archaeological sites of Fieri, the factories and markets of Korça, and the agricultural plant of Sukth. Tourists could enjoy sports facilities, and rent deckchairs and parasols on the beaches in front of the hotels.[31] Low-cost but quality consumerism was also mentioned as an attraction, with the adverts stressing the quality of the Albanian and international cuisine served in the hotels and the low price of consumer goods.[32] In the evenings, the advert promised, tourists could gather on the terrace of their hotel and enjoy ice creams, drinks, and cakes while listening to a live band playing Albanian traditional songs or Western pop music. A nightclub was also available. The trip was also presented as a chance to buy souvenirs in wood and copper, carpets, sheepskin coats and gloves, books, posters, and sweets, all at keen prices. As for the natural beauty of the country, the advert made much of the Adriatic and Ionian seas, the huge variety of vegetation, and the weather, with its "mild climate near the coasts but also eternally snow-capped alps."[33]

However, as pointed out by Derek Hall, the tourists were always accommodated in the same hotels run by Albturist, and were only allowed to visit a limited number of the country's attractions.[34] In general, the travel format proposed by the Association was similar to that of all the Albanian friendship associations, a fact not mentioned in the adverts.[35]

Dos and don'ts in Albania

After the split with the Soviet Union, Albania looked to tourism as both a propaganda tool and a source of hard currency, the latter being instrumental in guaranteeing the country's self-sufficiency. Evidently, the Albanian regime was adamant that the influx of tourists must not cause ideological contamination; the reverse in fact, as tourism existed to convince Western visitors that communist Albania was superior to their home countries.[36] These dynamics are visible in the brochures that the Swedish travellers received some weeks before departing. The tourists were provided with a detailed itinerary, which included a list of dos and don'ts, as specified in a letter sent to the Swedish travellers.[37] The letter claimed that the Albanians "do not view tourism as it is viewed in other places. They are not out to fleece tourists. Instead, they want to present their country to their guests."[38] The letter did not give examples of such "other places," but the reference was arguably aimed at its capitalist neighbours Italy and Greece, and its main competitor in tourism and politics, "deviationist" Yugoslavia. Contrary to the tourist models of those countries, Albanians were allegedly not interested in "profit," but wanted to "take care of their friends as guests," and expected "respect for Albanian traditions and culture" from them in return.[39] The letter introduced tourists to the philosophy of "friendship travel," but also to the formal rules and requirements which were effectively demanded of foreign visitors.

The tourists' first contact with Albania was at customs. In Tirana, a customs inspector would verify the decency of their clothes and hair and check their luggage for banned literature. Hair and dress codes were strict in Albania: beards and sideburns were considered reactionary, while long hair for men was considered "a capitalist tendency which aims at making the working class passive."[40] A barber was on hand at the airport to ensure they could comply on the spot. By contrast, "a neat well-grown moustache" was considered "acceptable."[41] Although it was not expressly forbidden to wear "extreme fashion," travellers were informed that they would "risk getting stared at if one does." Shorts and summer dresses were "normal on the beaches but not in the lunchrooms and on the daily excursions."[42] Women could instead wear long skirts or trousers, since "short skirts in Albania are considered an expression of women's exploitation" and "the Albanians are offended by short skirts."[43] Tourists could avoid outraging their hosts' sensibilities by changing on the spot or buying very cheap Albanian jeans on sale in the airport. They were also told it was forbidden to bring "religious, Trotskyite or revisionist literature" into the country, which after the Sino-Albanian split turned into a proscription against "pornographic and religious or anti-Albanian publications . . . [and] newspapers from Trotskyite or pro-Soviet/Chinese groups and parties."[44]

The limitations on dress and hair were the same for all foreigners visiting Albania. Philip Ward, the first to write an English travel guide to Albania in 1983, described his experience at the border with his clothes and hair in very similar terms.[45] Tourists who visited Albania with Ward's travel guide found it confiscated at the border, since Ward was accused of misrepresenting Albania.[46]

After the checks, the group would meet the Association's representative in Albania and the local guides, and travel with them to the hotel.[47] At the hotel, the guides would explain that the tourists could not move around independently, since the group had a collective visa. Apparently, the practice of issuing collective visas instead of individual ones for foreign tourists started after some of them, on an unspecified tour in 1969, had "behaved very badly," giving money and chewing gum to children.[48] The anecdote served to remind the group that it was obliged to stay together and that any deviation from the rules and customs was forbidden and offensive to their hosts. The guides also explained that photography was only permitted with their authorisation.[49]

By advertising the sunny beaches and warm sand, the Association explicitly invited along Swedish tourists who were not interested in anything else than swimming and relaxation. A letter assured the tourists that if, despite the activities available, one wanted "to stay on the beach to sunbath and swim" they could "definitely do so."[50] Group activities and individual relaxation were part of the mix offered to tourists across the Cold War socialist bloc.[51] Albania was no exception. Western tourists brought in hard currency, and each day trip represented additional income, so it was important for Albturist to fill the buses. The Association's ambition was to invite the tourists who were not members of the Association to join in the planned "purposeful leisure" activities without sounding insistent and without minimising their right to rest. The same was true of the preparatory meetings for each daily excursion. An internal Association document warned the guides to "remember that this kind of work must be voluntary. Many are not used to meetings and may experience it as a big constraint to gather in this kind of group. [Still] they often become interested and will get used to it later."[52] Handling the customers with care, in a respectful manner, would provide more revenue for Albturist and was a good advert for the Association and for the trips to come.

Between tourism and friendship

As noted by Paul Hollander, the favourable assessments of socialist countries by tourists were intended to discredit criticism and, by sheer repetition, lend plausibility to their narratives.[53] For this reason, one of the Association's major concerns was obtaining feedback from its customers, since travel reports were a means of propagandising in favour of the People's Republic of Albania, as well as being possibly the best means available to advertise the package tours to the public.[54] Therefore, the Association asked customers to submit travel accounts, and after each tour the board members invited the tourists to meet and discuss their travel experiences. The board and the tour leaders also met once a year to review the reports – in the presence of the Albanian Ambassador.

After 1980, the Association increased its propaganda efforts by publishing the bimonthly journal *Albanien och vi* ('Albania and us'). It informed the Swedish readership about the achievements of the regime and reprinted several positive travel accounts, with enthusiastic tourists talking about how Albania "is so incredibly beautiful" and inviting the readers to "see it with your own eyes."[55] The reports praised the lack of car traffic, the pleasantness of the hotel in Durrës and the beaches nearby, the high quality and the quantity of the meals served, and the enthusiasm encountered on the day trips to the war and liberation museums or to the archaeological sites. They all neatly echoed the canon established by the Association's tourist advert.

Just one unpublished travel report remains in the Association's archive: a collective report by a group who visited Albania in March 1977. The report was openly critical of the Albanian regime. Written in the form of a diary kept by one traveller, a woman who had visited Albania with a group of nineteen, it portrays how the trip was experienced differently by some of its participants, shifting between author's voice and observations by the other participants. It had four appendices, three of them written by one author each and the last one by two authors, which offered different views on many of the topics treated in the diary.

The diary began by narrating the confiscation of "a couple of books and posters we thought to leave as presents at the various study visits" as they entered the country.[56] The confiscation was a disappointment, because of the sincere intentions of the travellers. The author did not explicitly treat this episode as a source of concern, but still she mentioned it. The fact that one of the group had to get a shave was described in positive terms, since it was offered for free and thus demonstrated the allegedly unselfish nature of Albanian communism.

The author did not hide the impression that a meeting with a Turkish diplomat at the bar of the hotel left on the group. The diplomat claimed he was under constant surveillance by the Albanian military: "he says that the military check on him constantly. He thought that Albania was led by 2,000 families who are rich and have flashy lifestyles, while the broad masses are really poor. Enver [Hoxha], for example, has a car with six doors, he said. . . . Can it be true?"[57] After this encounter, the group felt "paranoid to see so many soldiers" in the streets, and they started to notice the Mercedes of the high-ranking members of the Labour Party of Albania. Later the group concluded that "to see the military here in Albania and e.g. in South America is not the same thing. In Albania there is a People's Army that stands for the peoples' interests (we hope)."[58] The explicit comparison of Albania with a South American military dictatorship is peculiar and raised far more questions than it answers.

While the report articulated a degree of trust in the Albanian regime, some of its features were explicitly said to be backward. This was particularly true of labour and human conditions. The majority of workers employed in a textile factory that the group visited turned out to be women, and the visitors noted that "the working environment was dangerous," even though their guide reassured them that "no big accident has happened."[59] The most ambiguous comments referred to a monument dedicated to a 15-year-old girl who had died doing voluntary work for the construction of the railway. " 'She died on duty'," the guide said, and the writer commented, "every kilometre in Albania shall have a monument in order to remember the struggle."[60]

The visitors suspected that the Albanian women were not as emancipated as was said. After the group noted that in the evenings no women were out on the streets, which were crowded exclusively with men, the author asked sarcastically whether they were "possibly at home to put babies to sleep and prepare the evening dinner."[61] The interactions between the guide and the group during the visit to a lipstick and perfume factory provided another example of the clash between utopian expectations and the Albanian realities:

> 'Why do you produce so many cosmetics?' asked one of the girls.
> The guide . . . answered shortly 'Don't you use lipstick and perfume?'
> 'No', answered the girls in chorus! We tried to explain that we see those goods as capitalist habits.[62]

As noted elsewhere, tourism can generate misunderstandings between the tourists and the hosts due to their different cultural backgrounds and experiences.[63] This case revealed a discrepancy between the wish of the host to highlight the country's development and the visitors' cultural perceptions. The Albanian asked in good faith about the Swedish visitors' use of cosmetics, since one of the goals of the Albanian regime (and that of communist countries in general) was to offer some degree of consumption, often presented as achievements of the revolutionary struggle. However, the Swedish feminists (including the ones touring Albania) had declared war on lipstick and perfume as capitalist habits, causing a mutual lack of understanding.

The sensitivities of the travellers were taken to a new level when the group engaged with the guides in a discussion about Albania's prison system. The Swedes noted that the prisoners "are still a part of the society and contribute to the construction of socialism." However, when the group asked whether it was true that homosexuality was punished by jail, the guides answered that "of course it is forbidden here in Albania," causing consternation.[64] "At first," the group noted, the guides "pretended that . . . [homosexuality] did not exist in Albania, but once we spoke about it they admitted that it existed. . . . They [the Albanians] distinguish two types of homosexuality. One part of them [the homosexuals] are sick and they are placed in mental hospitals. Those who do it 'on purpose' end up in prison."[65] The same was true of lesbianism, the existence of which was also denied in Albania. The author of the report noted "as usual, I sigh. A negation of women's sexuality."[66]

From the diary, it becomes clear that some participants were searching for a direct, emotional contact with the people, in order "to confront our own Socialist utopias . . . during our trip into this enigmatic land."[67] Such encounters could work both ways, and the few contacts the group had with the Albanian people were "strange," since the members of the group were perceived first and foremost as tourists by the local population. The gazes "were not critical. The eyes that met us were friendly, but it was a strange feeling anyway."[68] The guides only helped marginally during such encounters, insisting on supplying the tourists with "very detailed information that one finds uninteresting."[69] After a few days in Albania, the writer asked herself:

> Why do we always meet this distance between us and the local population? We came to Albania to meet Albanian culture and the Albanian population. Are the Albanians not prepared to receive us with open arms in their own everyday lives?[70]

The tension in encounters with a local culture and people is a recurring issue in tourism semiotics. According to Jonathan Culler, tourists tend to escape the stage set by the organisers, and, wishing for authenticity, try to connect directly with the everyday life of the place they are visiting.[71] Confronted with the "staged authenticity" set by the travel organisers, which made Albania "a kind of living museum," some of the Swedish tourists were simply bored.[72] By contrast, they reported with great enthusiasm on each and every unplanned incident that brought them into direct contact with the locals. The group was very disappointed when, on the last evening, the guides invited an orchestra to the hotel, since this prevented "the last chance to experience Albanian everyday life."[73] During the evening, the group counter-staged a sort of polite and liberating protest. One of the tourists went to the orchestra and hummed a traditional Swedish song,

> in less than 20 seconds those fantastic musicians understood the tune and we whipped around in a calm old-time dance. . . . All the hotel personnel came in and watched us, fascinated. Outside sat a gang of youngsters and they listened to the music and tried to catch a glance of us every time the doors opened. They did not come in. We will never know why.[74]

The 1977 travel diary represents well the desire for freedom and liberation. It openly points out which aspects the tourists appreciated and which they considered boring, disturbing, or shocking. By contrast, the four separate accounts appended to the diary show a tendency to see Albania through glasses tinted by ideology. Only one of them openly criticised the Albanian regime for its neglect of the living and working conditions of Albanian women. Its author had seen women working "in the sun and with primitive instruments," toiling in the factories they visited or cleaning the streets late at night.[75] The appendix suggests that the regime assigned a secondary role to women, referring to the socialist motto "Socialism cannot be built without women" on the one hand, and quoting the bold claim made by the tourist guide on the other: "In Albania, there is no private property – except for the housewife!"[76]

The other three appendices, by contrast, recollected some of the disturbing experiences witnessed during the trip. The second appendix, signed by a man and a woman, stated that for the group "it was difficult in the first days to see Albania in the right perspective, we did not see and understand the whole picture. We compared the little we saw with the corresponding information in Sweden without contextualizing."[77] Comparing Sweden with Albania on the basis of hard data must indeed have been disappointing: while Sweden had two preschool teachers for every ten children, Albania had two for every 25 to 30 children. However, the authors stated that "it is wrong to conclude that we have much better child-care in Sweden."[78] Albania, the travellers suggested, was safer than Sweden; in the former, a child could grow "in harmony" with the rest of society. Therefore, "society in Sweden should make it possible for the childcare centres to function with twenty-five to thirty children for every two teachers."[79] By idealising the construction of socialism in Albania, the authors concluded that Albania was far more just and equal than Sweden, which was presented negatively as a "technocracy" and an unsafe and dangerous place. Consequently, while realistic comparisons between Albania and Sweden would have painted a bleak picture of the last self-declared Stalinist country in Europe, ideological interpretations helped to preserve the initial fascination with socialist Albania against the backdrop of the leftist critique of their Swedish homeland.

Not all the appendices were so logically flawed. The author of the third appendix based the narrative on his own perceptions, rather than on hard data comparisons between Albania and Sweden, which would have highlighted Albania's evident backwardness and lack of development. A comparison between the capitalist countries and Albania led the author to conclude that "it is really evident that we live in a society based on competition where men are knocked out, and that [this] generates criminality, abuse of different kinds and other social problems."[80] This says much about the author's views on Sweden, but not on Albania. The author of the report admitted that the trip had opened his eyes to his homeland, causing him "a cultural shock": "it felt hard to come home to Sweden and compare [Albania] with the old usual world that suddenly felt alien and repulsive. The country's deficiencies and injustices became much more apparent."[81]

In absence of further unedited reports, it is difficult to say whether the criticism expressed in the 1977 diary had been a constant feature in the tourist feedback over the decade. Needless to say, such negative characteristics never appeared in the narratives of the members' journals, or, from 1980, in *Albanien och vi*. By contrast, criticism of Albania and the organised tours had featured since 1969 on the pages of major Swedish newspapers. The journalist Barbro Josephson, who had gone on the first tour to Albania in May 1969, was the first to publicly criticise Albania. In a series of articles in the national daily *Dagens Nyheter*, her picture of Albania had little trace by Marxist-Leninist propaganda or tourist adverts. Josephson's account was negative about her tourist experience as well as the situation in the country as a whole. She described the hotel portrayed so pompously in the tourist brochures as "a Russian building of Stalin's period," where water from the shower flooded the floor and the restaurant served abundant but tasteless food. Josephson reported the disapproving gazes of the locals at the miniskirts of the Swedes. She described the scene of a young boy hitting a tourist with stones. Finally, she denounced the police's confiscation of a tourist map. Josephson concluded that Albania was still cheaper than the Canary Islands, but tourists were not really as welcome there as the brochure said.[82]

In the summer of 1969, Josephson continued to write unfiltered descriptions of her travels in *Dagens Nyheter*. She pointed out that poverty in Albania existed in parallel to strictly guarded villas in central Tirana. Josephson also reported on the lack of freedom of mobility inside Albania. She pointed out that this problem was due to the lack of means of transportation, but also to the restrictive legislation. The Albanians she got to know needed to ask for special permission from the police even to travel outside their place of residence. Anyway, she was told, a 'proper' Albanian "would not dream of travelling abroad and therefore has no passport." From her experience she concluded that the hostility of the Albanian children towards tourists was a consequence of the regime's propaganda against its foreign enemies. Finally, she ridiculed the planned economy, stating that it was obvious to the eye of the most inexperienced tourist that Chinese imports were providing even the most basic consumer goods.[83] Unsurprisingly, Josephson was harshly criticised by the leftist newspaper *Tidsignal*, which launched an ad hominem attack, saying that her "divergent opinion was due completely to . . . [her] bourgeois prejudice . . . [her] negative attitude and other flaws in [her] personality."[84] Criticism of Albania and package tourism to the country continued to appear in Swedish newspapers.[85] At the same time, the number of tourists progressively diminished.

Political pilgrims, fellow travellers, bourgeois tourists

The Swedish-Albanian Association helped to propagandise a positive image of the People's Republic of Albania in Sweden until the fall of the communist regime. Organised travel to Albania was its main activity, one that could easily be sold as a leisure pursuit at the same time. In organising this propaganda effort, business motives counted as much as political ones, because planning several tours each year meant organising a business process involving multiple actors. For this reason, Albania was advertised as both a worker's and a tourist's paradise. This created different expectations from the two sets of travellers.

Paul Hollander has defined "political pilgrims" as those who travelled to the communist states in search of an ideological justification for the criticism of their homelands. The Swedish political pilgrims who organised travel to communist Albania considered it the realisation of their utopian dream of a just and ethical society, and chose to believe that it might be real. Steady in this belief, the "friends of Albania" became connoisseurs of Albanian politics, economy, history, and culture. Pål Steigan, leader of Norway's Workers' Communist Party, once pointed out that "no one else in Norway knows so much about Albania as we do."[86] He had a point, insofar as friendship associations and Albania itself had a virtual monopoly on information about Albania. As "experts," the political pilgrims wrote about Albania, held public conferences, and organised and advertised guided tours in cooperation with Albturist and the Albanian embassy.

The political pilgrims proselytised about their passion for Albania by welcoming "people with bourgeois opinions" to join them.[87] The Association exploited those who took part in these trips to legitimise the Albanian regime in the Western hemisphere, by making them background actors in the narrative of the construction of socialism in Albania. Consciously or not, by their presence in Albania they became fellow travellers of the Albanian regime.[88] Were all of them aware that the tours were political, and did they think this as the main reason to visit Albania? According to Silke Neunsinger and Iben Vyff, tourists in communist regimes were well aware that the tours were political, and this was how they experienced the trip, sympathising or criticising the state of things they met.[89] Nevertheless, as noted by Sune Bechmann Pedersen, rarely were politics staged as the main political attraction in these kinds of travel.[90] In the case analysed here, the leaders of the Association succeeded in raising interest in travelling to Albania by voluntarily blurring the distinction between political offer and tourist offer, between political "friend" and tourist customer.[91] Aiming to exploit diverging tourist desires, the Association promised emotional meetings with the people in their daily lives, together with visits to cultural and natural attractions, and the pleasures of socialism and sunny beaches.

Tourists did criticise the social conditions they encountered in Albania. Besides the unpublished travel report analysed in this chapter, open opposition to the Association's narrative was published in newspaper articles and in letters to the press. Some travellers cast about for ideological justifications for the situation in the country, while others complained that they could not "meet the people," or that the daily trips were exhausting or badly planned. We do not know how many expressed criticism, but, comparing the number of members of the Association and the number of travellers to Albania, it becomes evident that the majority of the tourists were not interested in becoming "friends of Albania."

It is doubtful that the "bourgeois" tourists consciously supported the Albanian regime, but it is also doubtful that they were convinced that their days in Albania were in any way similar to the ones spent by the local population. They paid in foreign currency for goods and tourist services – and luxury products (by Albanian standards) – and gave coins and bubble gum to poor children they met in the streets. Nevertheless, by taking part in the package tours, whatever their political or tourist expectations, they contributed to the propaganda designed to polish the image of a Stalinist regime at the international level.

Notes

1 This research has been generously financed by a postdoctoral scholarship from Åke Wibergs Stiftelse. Danresor, "Vi gjorde en telefonintervju med dom första gruppresenärerna nere i Albanien," *Dagens Nyheter*, May 25, 1969, 12.
2 Derek R. Hall, "Foreign Tourism Under Socialism: The Albanian 'Stalinist' Model," *Annals of Tourism Research* 11, no. 4 (1984): 539–55; Brunilda Liçaj and Armada Molla, "Peeping Tourist: A Case Study of the State Tourist Agency – AlbTourist," *9th Conference of the International Association for the History of Transport, Traffic and Mobility*, Berlin, Freie Universität, 2011.
3 Lorenz M. Lüthi, *The Sino-Soviet Split: Cold War in the Communist World* (Princeton: Princeton University Press, 2008), 201–9.
4 Elez Biberaj, *Albania and China: A Study of an Unequal Alliance* (Boulder: Westview, 1986), 63.

5 Michael David-Fox, "The Iron Curtain as Semipermeable Membrane: Origins and Demise of the Stalinist Superiority Complex," in *Cold War Crossings: International Travel and Exchange across the Soviet Bloc*, ed. Patryc Babiracki and Kenyon Zimmer (Arlington: University of Texas, 2014), 14–39; György Péteri, "Sites of Convergence: The USSR and Communist Eastern Europe at International Fairs Abroad and at Home," *Journal of Contemporary History* 47, no. 1 (2012): 3–12; Robert J. Alexander, *Maoism in the Developed World* (London: Praeger, 2001), 1–5; Anne Hedén, *Röd stjärna över Sverige: Folkrepubliken Kina som resurs i den svenska vänsterradikaliseringen under 1960- och 1970-talen* (Sekel: Lund, 2008).

6 Kjell Östberg, *1968 – när allt var i rörelse: Sextiotalsradikaliseringen och de sociala rörelserna* (Stockholm: Norstedts, 2002), 160; Hedén, *Röd stjärna över Sverige*, 18, 171, 240–41, 321; Kalle Holmqvist, "Den något mindre polemiken: SKP, KPML(r) och brytningen mellan Kina och Albanien," *Arbetarhistoria* 131, no. 3 (2009): 19–21.

7 Thomas Ekman Jørgensen, "Split or Reform? The Danish and Swedish CP's Facing the Post-Stalin Era," *EUI Working Chapter HEC*, no. 4 (2002): 45–47; See also Liçaj and Molla, "Peeping Tourist."

8 Birgitta Almgren, *Dröm och verklighet: Stellan Arvidson – kärleken, dikten, politiken* (Stockholm: Carlssons, 2016), 229–30; Hedén, *Röd stjärna över Sverige*, 143.

9 Stefan Hall, "Lidande, motstånd och våldets mekanik: Maskulina strategier i Gnistan 1969–1981" (MA diss., Stockholm University, 2013), 23–24.

10 Interimstyrelsens Verksamhetsberätelse, n.d., box 3638-1, Svensk–Albanska Förening-ens Arkiv (Swe–Alb), Swedish Workers' Movement Archive (Arab), Flemingsberg, Sweden. All translations are by the author.

11 Ibid.

12 Svensk–Albanska Föreningen, Aug – 70, August 1970, box 3638-1, Swe–Alb, Arab; Protokoll, Mariebergsgården, n.d., box 3638-1, Swe–Alb, Arab.

13 Protokoll, 8 July 1970, Box 3638-1, Swe–Alb, Arab.

14 Sune Bechmann Pedersen, "Eastbound Tourism in the Cold War: The History of the Swedish Communist Travel Agency Folkturist," *Journal of Tourism History* 10, no. 2 (2018). See also Elitza Stanoeva in this volume.

15 Protokoll, 8 July 1970, Box 3638-1, Swe–Alb, Arab.

16 Protokoll, 14 April 1970, Box 3638-1, Swe–Alb, Arab.

17 Verksamhetsberättelse från styrelsen 1971, p. 3, 1971, Box 3638-1, Swe–Alb, Arab.

18 Protokoll, 1 February 1971, Box 3638-1, Swe–Alb, Arab.

19 Verksamhetsberättelse 1972, 2, n.d., Box 3638-1, Swe–Alb, Arab.

20 Kort handledning för svensk – albanska föreningens gruppledare, n.d., 3638–11, Swe–Alb, Arab.

21 Nils Gösta Ekman and Josef Lundaahl, "Svensk student fängelsehotad för bibelspridn-ing i Albanien," *Svenska Dagbladet*, August 29, 1971, 7.

22 Svenska-Albanska Föreningens Verksamhetsberättelse 1976, 1, n.d., 3686–1, Swe–Alb, Arab.

23 The gradual split was reflected in the worsening relations between the pro-Chinese and pro-Albanian parties and associations in Sweden since April 1977. Hedén, *Röd stjärna över Sverige*, 239–41.

24 See Kallelse till föreningstyrelsemöte, 21–22 February 1981, 3686–2, Swe–Alb, Arab; see also Protokoll, 29 July 1980, 3686–2, Swe–Alb, Arab.

25 Protokoll, 6 February 1982, 3638–2, Swe–Alb, Arab.

26 Synpunkter på Albanienresan 23/6–8/7 1982, 1982, 3638–2, Swe–Alb, Arab.

27 Protokoll, 1 September 1984, 3638–2, Swe–Alb, Arab; See also Protokoll, 16 Novem-ber 1985, p. 2, 3638–2, Swe–Alb, Arab.

28 Anon, "Nöjda premiär-turister," *Albanien och vi*, no. 3 (1987): 3.

29 Anon, "Albanien: Inklämt och isolerat," *Aftonbladet*, May 4, 1974, 15.

30 Varför just Albanien, n.d., 3638–11, Swe–Alb, Arab; Badsemester Albanien, 1, n.d., 3638–11, Swe – Alb, Arab.
31 Vad du behöver veta innan du reser till Albanien, 2, n.d., 3638–11, Swe–Alb, Arab; see also Philip Ward, *Albania: A Travel Guide* (Cambridge: Oleander, 1983), 2.
32 Ibid.
33 Nu är du välkommen till Albanien! 3, n.d., 3638–11, Swe–Alb, Arab.
34 Hall, "Foreign Tourism Under Socialism," 549–52.
35 See, for example, Venskabsforeningen Danmark–Albanien, 3686–2, Swe–Alb, Arab.
36 Derek R. Hall, "Stalinism and Tourism: A Study of Albania and North Korea." *Annals of Tourism Research* 17, no. 1 (1990): 36–54.
37 Till våra albanienresenärer, n.d., 3686–11, Swe–Alb, Arab.
38 Ibid.
39 Vad du behöver veta innan du reser till Albanien, 1, n.d., 3686–11, Swe–Alb, Arab.
40 Ward, *Albania*, 9; Vad du behöver veta innan du reser till Albanien, 1, n.d., 3686–11, Swe–Alb, Arab.
41 Ibid.
42 Till våra albanienresenärer, n.d., 3686–11, Swe–Alb, Arab.
43 Upplysningar om Gnistans Studie- och Ferieresor till Albanien 1970, n.d., 3638–1, Swe–Alb, Arab.
44 Vad du behöver veta innan du reser till Albanien, n.d., 3686–11, Swe–Alb, Arab.
45 Ward, *Albania*, 9.
46 Hall, "Stalinism and Tourism," 45.
47 Kort handledning för svensk–albanska föreningens gruppledare, 1, 3638–1, Swe–Alb, Arab.
48 Anon, "Nytt studiematerial för Albanien–turister och andra vetgiriga," *Albanien och vi*, no. 2 (1980): 17.
49 Vad du behöver veta innan du reser till Albanien, 2, 3638–1, Swe–Alb, Arab.
50 Till våra albanienresenärer, n.d., 3638–11, Swe–Alb, Arab.
51 Adam T. Rosenbaum, "Leisure Travel and Real Existing Socialism: New Research on Tourism in the Soviet Union and Communist Eastern Europe," *Journal of Tourism History* 7, no. 1–2 (2015): 164.
52 Kort handledning för svensk–albanska föreningens gruppledare, 2, n.d., 3638–1, Swe–Alb, Arab.
53 Paul Hollander, *Political Pilgrims: Western Intellectuals in Search of the Good Society*, 4th ed. (New Brunswick: Transaction, 1998), 5.
54 Föreningsmöte, 16 October 1971, 3638–1, Swe–Alb, Arab. See also Reserapport från Albanien–gruppen 22/7–4/8 1978, 2, n.d., 3638–11, Swe–Alb, Arab.
55 Anon, 'Hur är det att vara turist i Albanien?' *Albanien och vi*, no. 2 (1980): 16–17. Anon, 'Se det med egna ögon,' *Albanien och vi*, no. 2 (1890): 24.
56 Vår resa till Albanien 1977, 3, n.d., 3686–11, Swe–Alb, Arab.
57 Vår resa till Albanien 1977, 17, n.d., 3686–11, Swe–Alb, Arab.
58 Vår resa till Albanien 1977, 18, n.d., 3686–11, Swe–Alb, Arab.
59 Vår resa till Albanien 1977, 6, n.d., 3686–11, Swe–Alb, Arab.
60 Vår resa till Albanien 1977, 7, n.d., 3686–11, Swe–Alb, Arab.
61 Vår resa till Albanien 1977, 6, n.d., 3686–11, Swe–Alb, Arab.
62 Vår resa till Albanien 1977, 18, n.d., 3686–11, Swe–Alb, Arab.
63 Rachel Applebaum, "A Test of Friendship. Soviet–Czechoslovak Tourism and the Prague Spring," in *The Socialist Sixties: Crossing Borders in the Second World*, ed. Anne E. Gorsuch and Diane P. Koenker (Bloomington: Indiana University Press, 2013), 213–34; see also Anne E. Gorsuch,"Time Travelers: Soviet Tourists to Eastern Europe," in *Turizm: The Russian and East European Tourist under Capitalism and Communism*, ed. Anne E. Gorsuch and Diane P. Koenker (Ithaca: Cornell University Press, 2006), 215–17.

154   *Francesco Zavatti*

64   Vår resa till Albanien 1977, 13, n.d., 3686–11, Swe–Alb, Arab.
65   Ibid.
66   Ibid.
67   Värdering av Albanienresan den 7–21/7 1978, 1, n.d., 3638–11, Swe–Alb, Arab; Vår resa till Albanien 1977, 19, n.d., 3686–11, Swe – Alb, Arab.
68   Värdering av Albanienresan den 7–21/7 1978, 4, n.d., 3638–11, Swe–Alb, Arab.
69   Ibid.
70   Vår resa till Albanien 1977, 19, n.d., 3686–11, Swe–Alb, Arab.
71   Jonathan Culler, "The Semiotics of Tourism," *American Journal of Semiotics* 1, nos. 1–2 (1981): 131.
72   Dean MacCannell, *The Tourist: A New Theory of the Leisure Class* (Berkeley & Los Angeles: University of California Press, 1999), 98–99.
73   Vår resa till Albanien 1977, 20, 1977, 3686–11, Swe–Alb, Arab.
74   Ibid.
75   Några tankar efter Albanienresa, 1977, 3686–11, Swe–Alb, Arab.
76   Ibid.
77   Personaliga tankar kring denna reserapport, 1977, 3686–11, Swe–Alb, Arab.
78   Ibid.
79   Ibid.
80   Ibid.
81   Ibid.
82   Barbro Jospehson, "Mur av ogillande möter Tirana-turist i kort-kort," *Dagens Nyheter*, May 28, 1969, 1–2.
83   Barbro Josephson, "Turist i Albanien har små chanser se mer än ytan," *Dagens Nyheter*, May 30, 1969, 11.
84   Barbro Josephson, "Kvinnlig eftertanke om albansk turism," *Dagens Nyheter*, June 11, 1969, 30.
85   Anon, "Bra om Spanien i Kvällsöppet," *Expressen*, October 1, 1975, 15; Maria-Pia Boethius, "Givande men följdfrågor saknades," *Expressen*, October 1, 1975, 28; Svensk–Albanska föreningens verksamhet i Helsingborg 1979–1980. 3686/8, Swe–Alb, Arab.
86   Pål Steigan, *På den himmelske freds plass: Om ml-bevegelsen i Norge* (Olso: Aschehoug, 1985), 169.
87   Tal, 1, 7 October 1978, 3686–1, Swe–Alb, Arab. The same trend was present in the Norwegian–Albanian Association; see Jon Rognlien and Nik Brandal, *Den Store Ml-boka: Bonuskapitler* (Oslo: Kagge Forlag, 2009), 65.
88   David Caute, *The Fellow-Travellers: Intellectual Friends of Communism* (New Haven: Yale University Press, 1988). On narratives of communism in Sweden, see Kristian Gerner, "Kommunistiska berättelser i Sverige – en fråga om tro," in *Rysk spegel: Svenska berättelser om Sovjetunionen – och om Sverige*, ed. Kristian Gerner and Klas-Göran Karlsson (Lund: Nordic Academic Press, 2016), 17–30.
89   Silke Neusinger and Iben Vyff, "Indledning: Turisme, solidaritet og politik," *Arbejderhistorie/Arbetarhistoria* 137, no. 1 (2011): 5.
90   Sune Bechmann Pedersen, "Politisk turisme og turismepolitik: Fornøjelsesrejser til Østeuropa under den kolde krig," in *Turismhistoria i Norden*, ed. Wiebke Kolbe (Uppsala: Kungl. Gustav Adolfs Akademien, 2018), 59–70.
91   In Norway, the pro-Albanian party chose which members could join the tours, and required that they introduce themselves as "tourists". See SKs arbeidsutvalg, "Direktiv til alle Lagsstyrer om hvordan partimedlemmene skal oppføre seg på Albania–turene i sommer," May 24, 1977, AKP online arkiv, accessed October 3, 2017, www.akp.no/ml-historie/pdf/direktiver_og_meldinger/albania_240577.PDF. They were to join the Norwegian–Albanian Association after the holiday. See Kristine Mollø-Christensen, "Albania er ingen dominobrikke," *Revolusjon*, no. 1 (1990).

References

Alexander, Robert J. *Maoism in the Developed World*. London: Praeger, 2001.

Almgren, Birgitta. *Dröm och verklighet. Stellan Arvidson: Kärleken, dikten, politiken.* Stockholm: Carlssons, 2016.

Anon. "Albanien: Inklämt och isolerat." *Aftonbladet*, May 4, 1974, 15.

———. "Bra om Spanien i Kvällsöppet." *Expressen*, October 1, 1975, 15.

———. "Nytt studiematerial för Albanien – turister och andra vetgiriga." *Albanien och vi*, no. 2 (1980): 17.

———. "Nöjda premiär-turister." *Albanien och vi*, no. 3 (1987): 3.

Applebaum, Rachel. "A Test of Friendship: Soviet–Czechoslovak Tourism and the Prague Spring." In *The Socialist Sixties: Crossing Borders in the Second World*, edited by Anne E. Gorsuch and Diane P. Koenker, 213–34. Bloomington: Indiana University Press, 2013.

Bechmann Pedersen, Sune. "Eastbound Tourism in the Cold War: The History of the Swedish Communist Travel Agency Folkturist." *Journal of Tourism History* 10, no. 2 (2018): 130–45.

———. "Politisk turisme og turismepolitik: Fornøjelsesrejser til Østeuropa under den kolde krig." In *Turismhistoria i Norden*, edited by Wiebke Kolbe, 59–70. Uppsala: Kungl. Gustav Adolfs Akademien, 2018.

Biberaj, Elez. *Albania and China: A Study of an Unequal Alliance*. Boulder: Westview Press, 1986.

Boethius, Maria-Pia. "Givande men följdfrågor saknades." *Expressen*, October 1, 1975, 28.

Caute, David. *The Fellow – Travellers: Intellectual Friends of Communism*. New Haven: Yale University Press, 1988.

Culler, Jonathan. "The Semiotics of Tourism." *The American Journal of Semiotics* 1, no. 1–2 (1981): 127–40.

David-Fox, Michael. "The Iron Curtain as Semipermeable Membrane: Origins and Demise of the Stalinist Superiority Complex." In *Cold War Crossings: International Travel and Exchange across the Soviet Bloc*, edited by Patryc Babiracki and Kenyon Zimmer, 14–39. Arlington: Texas A&M University Press, 2014.

Ekman Jørgensen, Thomas. "Split or Reform? The Danish and Swedish CP's Facing the Post-Stalin Era." *EUI Working Chapter HEC* 4 (2002): 45–75.

Ekman, Nils Gösta, and Josef Lundaahl. "Svensk student fängelsehotad för bibelspridning i Albanien." *Svenska Dagbladet*, August 29, 1971, 7.

Gerner, Kristian. "Kommunistiska berättelser i Sverige – En fråga om tro." In *Rysk spegel: Svenska berättelser om Sovjetunionen – och om Sverige*, edited by Kristian Gerner and Klas-Göran Karlsson, 17–52. Lund: Nordic Academic Press, 2016.

Gorsuch, Anne E. "Time Travelers: Soviet Tourists to Eastern Europe." In *Turizm: The Russian and East European Tourist under Capitalism and Communism*, edited by Anne E. Gorsuch and Diane P. Koenker, 205–26. Ithaca: Cornell University Press, 2006.

Hall, Derek R. "Foreign Tourism Under Socialism: The Albanian 'Stalinist' Model." *Annals of Tourism Research* 11, no. 4 (1984): 539–55.

———. "Stalinism and Tourism: A Study of Albania and North Korea." *Annals of Tourism Research* 17, no. 1 (1990): 36–54.

Hall, Stefan. "Lidande, motstånd och våldets mekanik: Maskulina strategier i Gnistan 1969–1981." MA diss., Stockholm University, 2013.

Hedén, Anne. *Röd stjärna över Sverige: Folkrepubliken Kina som resurs i den svenska vänsterradikaliseringen under 1960- och 1970-talen*. Sekel: Lund, 2008.

———. "Vittnen i vänsterkant." *Sydsvenskan*. Accessed September 14, 2009. www.syds venskan.se/2009-09-14/vittnen–i–vansterkant.

Hollander, Paul. *Political Pilgrims: Western Intellectuals in Search of the Good Society*. 4th ed. New Brunswick: Transaction, 1998.

Holmqvist, Kalle. "Den något mindre polemiken: SKP, KPML(r) och brytningen mellan Kina och Albanien." *Arbetarhistoria* 131, no. 3 (2009): 18–22.

Josephson, Barbro. "Mur av ogillande möter Tirana-turist i kort-kort." *Dagens Nyheter*, May 28, 1969, 1–2.

———. "Turist i Albanien har små chanser att se mer än ytan." *Dagens Nyheter*, May 30, 1969, 11.

———. "Kvinnlig eftertanke om albansk turism." *Dagens Nyheter*, June 11, 1969, 30.

Liçaj, Brunilda, and Armada Molla. "Peeping Tourist: A Case Study of the State Tourist Agency – AlbTourist." *9th Conference of the International Association for the History of Transport, Traffic and Mobility*, Berlin, Freie Universität, 2011.

Lüthi, Lorenz M. *The Sino-Soviet Split: Cold War in the Communist World*. Princeton: Princeton University Press, 2008.

MacCannell, Dean. *The Tourist: A New Theory of the Leisure Class*. Berkeley & Los Angeles: University of California Press, 1999.

Mollø–Christensen, Kristine. "Albania er Ingen Dominobrikke." *Revolusjon*, no. 1 (1990).

Neusinger, Silke, and Iben Vyff. 'Indledning: Turisme, Solidaritet og Politik.' *Arbejderhistorie/Arbetarhistoria* 137, no. 1 (2011): 3–6.

Nilson, Per. "En resa till Shqipëria – Örnarnas land (maj 1987)." Accessed June 1, 2016. http://perenn.com/en–resa–till–shqip–ria–rnarnas–land–maj–1987.

Péteri, György. "Sites of Convergence: The USSR and Communist Eastern Europe at International Fairs Abroad and at Home." *Journal of Contemporary History* 47, no. 1 (2012): 3–12.

Rognlien, Jon, and Nik Brandal. *Den Store Ml-boka: Bonuskapitler*. Oslo: Kagge Forlag, 2009.

Rosenbaum, Adam T. "Leisure Travel and Real Existing Socialism: New Research on Tourism in the Soviet Union and Communist Eastern Europe." *Journal of Tourism History* 7, no. 1–2 (2015): 157–76.

SKs arbeidsutvalg. "Direktiv til alle Lagsstyrer om hvordan partimedlemmene skal oppføre seg på Albania–turene i sommer." May 24, 1977, AKP arkiv. Accessed October 3, 2017. www.akp.no/ml-historie/pdf/direktiver_og_meldinger/albania_240577.PDF

Steigan, Pål. *På den himmelske freds plass: Om ml-bevegelsen i Norge*. Olso: Aschehoug, 1985.

Ward, Philip. *Albania: A Travel Guide*. Cambridge: Oleander, 1983.

Östberg, Kjell. *1968 – När allt var i rörelse: Sextiotalsradikaliseringen och de sociala rörelserna*. Stockholm: Norstedts, 2002.

Part III

The politics of tourism during the Cold War

8 Playing the tourism card
Yugoslavia, advertising, and the Euro-Atlantic tourism network in the early Cold War

Igor Tchoukarine

In the Cold War, Yugoslavia straddled the divide between East and West in a unique way. Led by the charismatic Tito, the country differentiated itself on the global stage by its distinct form of socialism, its engagement in the non-aligned movement, and labour and mobility policies that allowed its citizens to work abroad. Tourism similarly embodied the country's singular positioning, as numerous tourists from both sides of the Iron Curtain spent their holidays in Yugoslavia. The country's transformation into a worldwide tourist destination in the 1950s and early 1960s did not happen in a vacuum, though. It was rooted in growing professional and transnational networks of tourist and advertising experts, who played a significant role in shaping the tourism industry's practices and norms. Moreover – and surprisingly for a socialist country – American media associated Yugoslavia with traditional Western European destinations as early as 1952.

This chapter explores these dynamics, focusing on Yugoslavia's participation in the European Travel Commission (ETC) advertising campaigns in the US from 1952 to 1963 and its collaboration with the Organisation for European Economic Cooperation (OEEC) Committee for Tourism. These organisations, because of their close links with the Marshall Plan, with governments in Europe and North America, and with global organisations involved in tourism, offer compelling evidence of how the norms of tourism and mobility circulated during the Cold War. I would argue that they were Cold War embodiments of the intersection between new economic and political concerns, social and cultural values, and technological changes in the field of tourism. I also examine the 1958 US trade mission in Yugoslavia's publicity-driven recommendations, the trade mission's reception in the country, and the training of Yugoslav tourism experts in Western Europe in the early 1960s. As I demonstrate, Yugoslavia's tourist trajectory was both informed by and transcended the development of tourism in Europe, as practices and policies in the field were initiated, implemented, and revamped by a vast array of actors (international organisations, travel agencies, individuals, and so on). At the same time, tourism closely intersected with the country's geopolitical interests, hence the assertion by Yugoslav diplomat Vladimir Velebit that the country had "to play the tourism card."

Organising and advertising tourism: all eyes on America

The impact on tourism of the ETC and the OEEC (and from 1961 onwards, its successor the Organisation for Economic Cooperation and Development, or OECD) – as well as the history of the OEEC/OECD in general – remains relatively unexplored.[1] As an intergovernmental organisation, the OEEC in fact had no prerogative to establish binding regulations, perhaps explaining why it has been overlooked by scholars; rather, it sought to build consensus and widen its influence by soft-power persuasion. Likewise, from its inception in December 1948, the OEEC's Committee for Tourism had no real decision-making power, and was reliant on member states' willingness to implement its recommendations to further trade liberalisation, conduct research on tourism-related issues, and develop tourism. This did not preclude the OEEC from recognising the role that tourism could play in reducing the dollar gap and European countries' balance of payments, and the OEEC established the Committee for Tourism to further this objective.

Driven by similar free-trade principles, the ETC was established as the European Regional Commission of the International Union of Official Travel Organisations (IUOTO) by IUOTO's 14th general assembly in Norway in June 1948. The 47 countries that participated in the congress unanimously agreed on the importance of American tourism for European economic recovery, and a select committee of the 17 countries involved in the Marshall Plan decided to join forces on a collective advertising campaign in the US to promote European tourist destinations.[2] The first ETC publicity campaign in America, launched in the summer of 1949, had the enthusiastic support of American and French officials.[3] French tourism officials predicted the ETC would contribute significantly to member states' prosperity and the Marshall Plan's success.[4]

In the early Cold War, the 17 national tourism boards that comprised the ETC thus chose to concentrate on collective advertising campaigns in the US in order to attract more American tourists to Europe. In 1951, Finland, Spain, and Yugoslavia joined the ETC, a move eagerly supported in the West. In December of that year, the ETC President Arthur Haulot sent a report to the OEEC arguing that this enlargement would show Americans "a genuine desire for European cooperation," and that the joint publicity campaign "would be even more forceful, since it would give the American public practical proof that the whole of Western Europe is united in this campaign."[5] By lumping Yugoslavia in with Finland and Spain, Haulot's comments suggest that Yugoslavia was considered the crucial periphery that could well be considered Western, and a site whose political significance continued to grow in the early stages of the Cold War.[6] Similarly, the ETC's publicity campaigns never strayed far from the agenda set by political considerations. In March 1949, ETC representatives discussed the Marshall Plan in a plenary session, and concluded that "strong tourism is an efficient and important way to protect wide population strata . . . of European countries from the threat of communism."[7] Another of the Committee for Tourism's early tasks was to coordinate the efforts of the OEEC, the ETC, and the Economic Cooperation Administration (the agency that administered the Marshall Plan).[8] In fact, there was a clear symbiosis between the Committee for Tourism and the ETC: their meetings were held a day apart in Paris at the Château de la Muette, and their leadership overlapped considerably, especially when the ETC's first president Henry Ingrand became chairman of the Committee for Tourism in 1949.

In broad terms, the Committee for Tourism's and the ETC's goals were to develop transatlantic and intra-European tourism in the aftermath of the Second World War, and to diminish barriers to the free movement of people. To this end, the committee was commissioned to analyse the visa and passport regimes of its member states, to look at transatlantic transport, and to encourage the private sector as well as other international organisations to contribute to the tourism drive. Dealing with these issues placed the committee at the heart of the vast structuring and regulation of tourism in the 1950s. Both it and the ETC shared a vision of tourism's role: to rectify Europe's balance of payments, to support the reconstruction of its economies, to encourage European integration, to promote peace and understanding under the aegis of American and Western European leadership, and to support freedom of movement for tourists. There was a growing consensus among tourism experts and stakeholders that visas should be abolished. In April 1950, the Council of the OEEC recommended that its members "examine as quickly as possible the possibility of taking all necessary measures to abolish the visa requirement for all member state nationals."[9] In conjunction with this venture, which bore fruit in the early 1950s, intergovernmental cooperation led to the creation of zones where only a simple identity card was required to cross borders. Mobility regimes free of passport and visa requirements were implemented among Benelux countries; among Belgium, France, Luxembourg, and Switzerland; among Scandinavian countries; and between Ireland and the UK. These policies were welcomed by both the Committee for Tourism and ETC President Haulot, whose stated goal was the "total liberation" from obstacles to travel.[10] To this end, the Committee for Tourism predicted in late 1954 that passports would cease to exist once visas were eliminated among all member states.[11] Though this never came to pass, the committee did succeed in eliminating the caps on foreign currency that tourists could carry in the fifties, and the committee's general significance was upheld by the impressive growth of tourism.

These favourable circumstances led the Committee for Tourism to spearhead a series of studies on transportation, social tourism, tourism infrastructure, tourist destinations and length of stay, and visa regulations, as well as to collate and compare member states' tourism statistics. Here Yugoslavia in the early 1950s stood out from other member countries for its strict restrictions on mobility and its visa requirements for travelling into or out of the country. In 1954, Haulot laconically said of the Yugoslav visa regime that "everything [remained] to be done," but advised that on the settlement of the Trieste question Yugoslavia "should now seek the means to be deployed to ease foreigners' access to its territory."[12] Even though Yugoslavia remained an authoritarian regime with strict mobility policies and required foreigners to apply for entry visas, the country's decision to join the ETC at the end of 1951 dovetailed with other similar initiatives: Yugoslavia was also involved in many tourism-related projects with South Eastern and Western

European countries, as well as with international organisations and American marketing consultants. In 1953, the Yugoslav travel agency Putnik contacted a American public relations firm, Hal Leyshon & Associates, in order to promote tourism in Yugoslavia.[13] As the historian Patrick Patterson explains, from the mid-1950s Yugoslav marketing experts looked West for know-how and marketing strategies as well as to broaden their network.[14] "Even at this comparatively early stage, Western ways of thinking about advertising and marketing had begun to leave a deep imprint in the mentality of the Yugoslav [tourism] profession."[15] In the immediate aftermath of the Balkan Pact of 1953, the Tourism Association of Yugoslavia (Turistički Savez Jugoslavije, or TSJ) planned to work with Greece and Turkey in order to attract foreign tourists, and Americans in particular. Practical considerations prompted further collaboration. Marketing in America required huge investment, and the TSJ concluded that "Greece–Turkey–Yugoslavia" tours would be best promoted by a shared tourism office in New York.[16] In 1955, Yugoslavia hosted in Dubrovnik the 29th Congress of the International Federation of Travel Agencies (FIAV, the European counterpart of ASTA, the American Society of Travel Agents), which attracted representatives from 34 countries and numerous international organisations, including the ETC. In a moment of sheer optimism the Yugoslav political authorities proclaimed that "Yugoslavia has been at the service of international tourism" since 1950.[17] These initiatives speak to the increasingly more complex interconnectedness of the tourism industry in the postwar period. In this extent, we can best understand Yugoslavia's tourist identity as the result primarily of transnational processes, tourism experts and advertising, and international organisations such as the ETC.

ETC joint publicity campaigns in the US

"Travelling is learning how to live better," and Americans are visiting Europe and other places in the world in greater numbers than ever, proclaimed a 1953 cover story in *Tide*, a magazine that specialised in sales and advertising. This kind of article – even though it only briefly mentioned the ETC and its motto "Understanding through travel is the passport for peace" – was part of the ETC's considerable public relations effort to bring more Americans to Europe and publicise its work among American advertising and travel specialists.[18] This article, and many others like it, indicated the transnational dimension to cultural diplomacy and tourism promotion after the Second World War. Like the ETC, governments and travel agencies across Europe stressed the purposeful nature of tourism and looked at the American market with great interest.[19] For instance, in 1945 the French minister of public works requested funds for the immediate "reorganising and reopening" the French pre-war network of information and travel agencies, adding that the largest budget was for the New York office.[20]

In the early 1950s, the ETC launched a series of initiatives to promote Europe that included travelling exhibitions, window displays in New York's Rockefeller Plaza, radio broadcasts, short films, and brochures and calendars of events and festivals in all the member states. Adverts were placed in the leading newspapers, popular magazines, trade magazines, and foreign-language newspapers published in the US. The ETC also encouraged Americans to travel to Europe in the low season, emphasising the "lowest possible fares" and "wide range of hotel accommodations" and cultural activities that would be available.

Published by the ETC's publicity service and *This Week* magazine, the 1952 bestselling guidebook *Travel Key to Europe* is illustrative.[21] It offered the American public descriptions and travel advice for all 21 ETC member countries (including Yugoslavia). Not surprisingly, given the Cold War context, the journalist John Gunther wrote in the guidebook's preface that "We, the democrats of the world, *can* travel. There are millions upon millions of human beings behind the Iron Curtain who cannot."[22] By including Yugoslavia while obliterating the Soviet bloc from the map of Europe, it was an early example of Yugoslavia's inclusion among the most active global tourism actors, and highlighted its atypical geopolitical position. Using the standard language of tourism, Yugoslavia was praised for the hospitality of its people, the "startling scenic contrasts, sheer cliffs and turreted castles," and the visible "memories of an ancient world – of Rome and the Middle Ages" – and not a word about socialism.[23] This was no surprise given that in the early 1950s Yugoslavia was relatively unknown except for its association with communism. As Patterson notes, Eastern European tourist guides carefully calibrated their descriptions of Yugoslavia's topography and scenic and industrial landscapes so as to de-emphasise, for the citizens of the Soviet bloc, the unique aspects of Yugoslavia's socialist market economy and relative cultural and societal openness. In Eastern Europe, the beauties of Yugoslavia were conveniently emphasised over its socioeconomic and ideological otherness.[24] Pieces promoting and providing information about Yugoslavia were also disseminated through Yugoslavia's Information Office in New York. In 1951, it put together a brochure for the American public that used visual and textual narratives along these lines to promote the country (Yugoslavia's openness, religious freedom, desire for independence, beautiful landscape, and remarkable folklore), but that also stressed its modern and industrial development.[25]

The ETC did not neglect the public relations element to its campaigns. Thanks to its broad network and the receptivity of American journalists, writers, travel agents, and publishers, the ETC was often able to publicise its activities at little or no cost. From the outset, the ETC reminded national tourism organisations that joint advertising campaigns could achieve results that would be difficult to attain by other means.[26] The ETC structure was especially advantageous to small countries such as Yugoslavia, for which an independent marketing campaign in America would eat up a substantial part of any tourism budget, and proportionally more than larger countries could spend. Of the 1956 ETC publicity budget of $263,000, France provided $48,000, Italy $45,000, Great Britain $41,250, Germany $22,500, and Switzerland $15,000, accounting for 65 per cent of the total. In 1956, Yugoslavia contributed $2,500 to the ETC joint publicity budget, matching Finland's and Turkey's contributions, but eclipsing those of Iceland, Luxemburg, and Monaco. Compared to the French, Italian, British, German, and Swiss appropriations, and even Spain's contribution of $8,000, the Yugoslav contribution appears modest, but it was not atypical for a country of its size and location (members' contributions were calculated according to several factors, including its location, its population, and the number of American tourists who visited).[27] In 1961, Yugoslavia contributed 1.28 per cent (or $4,000) of the budget, and was slated to allocate 1.6 per cent (or $5,000) in 1962.[28]

From the outset, the issue of unequal contributions to the joint publicity budget sparked acrimonious debates. It even risked jeopardising the whole project during the OEEC's transformation into the OECD in 1961–62, when major contributors such as France, Switzerland, Austria, Portugal, and Spain challenged the efficiency of the ETC's operations. For all other OECD members, and especially for less frequented tourist destinations such as Finland, Denmark, Luxemburg, and Yugoslavia, the status quo was highly beneficial, and they openly supported the continuation of joint publicity campaigns, stressing their usefulness, efficiency, and necessity.[29] For Yugoslavia the advantage of being associated with these advertising campaigns was significant, as attested by the Yugoslav representatives' request in 1962 not to abandon the "new elements of tourist attraction" for "countries newly entered in tourism activities."[30]

Although Yugoslavia's participation (financial and otherwise) in the ETC was modest, it became associated in the American media with traditional Western European destinations – an illustration of Yugoslavia's early awareness of the need to compete and collaborate with major tourist destinations in order to secure a greater share of the American market and to bolster its reputation among Americans. Because it was not realistic to promote 21 countries in a single full-page advert, ETC advertising often showed member states' flags and featured a rotating cluster of countries. For example, in 1955 and 1956, ETC adverts in the travel magazine *Holiday* promoted Yugoslavia alongside Finland, the Netherlands, Turkey, the United Kingdom, Denmark, and Norway.[31] In addition to carrying numerous ETC adverts, the prominent American travel magazines *Holiday* and *Travel* published articles on Yugoslavia (five and seven respectively between 1950 and 1962). In 1953, *Travel* published an article on Yugoslavia that featured Dubrovnik as the prime place to visit. It described the country in positive terms, acknowledging that getting there might be difficult, but the trip was well worth the effort. Relying on common tropes, the article made much of the country's beauty and simplicity. Perhaps sensing that American readers needed to be reassured about their reception, the author wrote, "Unlike many European cities, Dubrovnik apparently harbors no anti-American sentiment. To the contrary, they exhibit considerable friendly curiosity," adding that "the tourist is not frequently reminded that Yugoslavia is a Communist country" with locals freely "condemning the regime in almost any language."[32] Given the tone, it is very unlikely the article was a direct product of Yugoslavia's promotional efforts in the US, but its publication – and its positive view of Yugoslavia – speak volumes about the integration of Yugoslavia into American travel-related media. A newspaper feature such as this would have been an indirect consequence of Yugoslavia's participation in the ETC campaigns, which were run under the aegis of the New York ETC coordination committee by the marketing firms Caples & Co and later Donald N. Martin & Co.

While *Holiday* and *Travel* covered both domestic and global travel, it is telling that they published more articles about Yugoslavia (and, indeed, other socialist countries) after 1953. Interest in Yugoslavia grew among American tourists, as letters to the ETC offices confirm. Between 1 January and 30 April 1953, the ETC offices in the US received 529 letters requesting information about Yugoslavia (by comparison, requests about Spain and France were 1,595 and 2,527 respectively).[33] Out of a total of 37,534 information requests in 1955, the ETC received 797 about Yugoslavia, compared to 2,458 about Spain and 3,628 about France.[34] In 1956, the ETC forwarded 46,418 requests to member tourist bureaus, 1,109 of them to Yugoslavia.[35] In 1957, the ETC redirected 50,078 requests, 1,040 to Yugoslavia as against 2,161 to Spain and 5,412 to France.[36] The number of Americans visiting Yugoslavia was modest, but grew steadily to reach almost 250,000 in 1970. In 1965, Yugoslavia welcomed more American tourists than Poland, Czechoslovakia, Hungary, Bulgaria, Romania, and the USSR combined (86,822 versus 85,594).[37] The benefits of being associated with the ETC's campaigns were plain, and tourism experts spearheaded further collaborations.

Transcending political borders: tourism networks

French tourism experts recruited American advertising agencies for the complex business of branding post-war France "as old yet modern, refined yet not too intimidating" in the American media.[38] In the tourism industry, as elsewhere in Europe, American advertising expertise was in great demand. The American firm of J. Walter Thompson, for instance, led the effort in the 1950s to improve public perceptions of the Marshall Plan and to revamp NATO's image among Europeans.[39] Similarly, from the mid-1950s, Yugoslav advertising experts looked to the West for advertising know-how, marketing strategies, and wider consumer networks. In this respect, Patterson believes that "Yugoslav advertising did in many ways represent a truly radical departure from the way things were done in most other places across Eastern Europe and the Soviet Union."[40] Unsurprisingly, advertising was deemed essential to tourism, and thus Yugoslavia solicited American expertise.

Following two US trade missions to Yugoslavia in 1956 and 1957, a third followed between 16 August and 26 September 1958, coinciding with the Belgrade Technical Fair and the Zagreb International Trade Fair, which attracted more than 25 countries (including the US and socialist states). The Zagreb newspaper *Vjesnik* declared that the 1958 trade mission would allow its members "to exchange ideas and create personal contacts with those who worked in commerce and tourism."[41] Sure enough, the mission's itinerary included Belgrade, Skopje, Titograd, Dubrovnik, Karlovac, Maribor, Ljubljana, Bled, Opatija, Rijeka, Zagreb, and Split, and allowed members – including the tourism and marketing experts Robert C. Gordon of *Time* magazine and Fred Wittner from Fred Wittner Advertising – to meet hundreds of Yugoslav professionals. Gordon and Wittner opined that Yugoslavia had "infinite possibilities to increase its tourism from the US." Drawing parallels between Yugoslavia and Spain, they emphasised that the latter "had only a handful of American tourists in 1947 and which, by a small but consistent and skillfully conducted advertising program . . . developed to the point where US tourists today provide the major source of dollar income for Spain."[42] The two maintained that the task ahead for Yugoslavia was exactly the same as Spain's had been:

> To build an image – a mind picture – of today's Yugoslavia: a country teeming with energy and new factories, a country richly endowed with natural beauties and which, although Communist in ideology, has dedicated itself to sharing the American traditions of independence and friendliness with other peoples.[43]

The local Yugoslav press seemingly adopted these views. The Titograd newspaper *Pobeda* reminded its readers that if "Spain has transformed old castles into hotels, the old village of Sveti Stefan could become a major attraction on the Adriatic coast as soon as Americans would know this site's accommodation options, its charm, and its moderate prices."[44] To this end, Gordon and Wittner recommended that Yugoslavia move its tourism office to the ground floor of its New York building (from the 22nd floor), then add window displays and "at least one American [employee], who is completely familiar with American selling techniques." They also considered it important for Yugoslavia to collaborate with leading US travel agents and tour operators. Americans, they advised, had to understand "that Yugoslavia is *not* behind the Iron Curtain . . . that it is a nation devoted to a philosophy of internal growth and national independence . . . that though it is definitely a Communist country, Americans are free to travel as they please."[45]

In their attempts to shape the Yugoslav image abroad, these experts acted as mediators, helping showcase a communist country located in the Balkans (and thus laden with negative connotations) as a fashionable and respectable travel destination. Leveraging his network, Wittner put numerous Yugoslav professionals in touch with American experts, while maintaining a detailed correspondence with the marketing manager of Yugoslavia Export.[46] In the magazine *Pregled*, published by the Yugoslav American Embassy's Information Services unit, Gordon and Wittner willingly accepted this mediating role, writing, "Yugoslavia can rely on us to pass on to our friends and trade partners . . . the image of an authentic Yugoslavia, which satisfies tourists during their summer vacations in a way that most other countries do not."[47] As the last quotation reveals, the increasingly interconnected world of tourism favoured the creation of networks and spaces. This was a time of multiple exchanges and tour projects between multiple countries on both sides of the Iron Curtain: the Soviet Union, its satellites, America, the Western Europe, neutral states, Yugoslavia.[48] Despite the numerous administrative and political obstacles generated by the Cold War, there were multiple ways for individuals and organisations to establish and cultivate professional contacts. Yugoslavia quickly understood the advantages of foreign training for its tourism industry workers, and between 1952 and 1967 sent almost 6,400 Yugoslavs to receive technical training in Western countries.[49] In the same period, Yugoslavia benefitted from the support of the OEEC (and later OECD), the UN, the US, and Western European government institutions such as the French Tourism Commissariat for professional training programmes. Among others, four groups of Yugoslav trainees (14 individuals in total), with interests ranging from tourism architecture to kitchen management and tourism advertising, studied in France and Switzerland in the early 1960s; their personnel files reveal their Western orientation, the interconnectedness of the professional tourism networks, and the importance Yugoslavia placed on playing the tourism card in a shifting post-war order.

The trainees' personnel files (held in the French National Archives) reveal that Yugoslavia's professional ties with the West were already well established by the early 1950s (roughly the time Yugoslavia started to collaborate with the ETC). Eleven of the trainees were in their late 30s when they studied in France, and most were already recognised specialists in their field. Their travel experiences were diverse, but focused on Western Europe; only one had not travelled abroad prior to his traineeship in France. Between 1951 and 1960 (and especially between 1954 and 1960), the other trainees had travelled for professional reasons to France, Italy, Switzerland, Austria, West Germany, and Belgium. By contrast, one had travelled to the USSR, and two others had visited Hungary and Czechoslovakia in 1946 and 1947. Janez Planina, who trained in France in 1961, had previously travelled to England, Italy, the Netherlands, Austria, and Switzerland, where he participated, in 1960, in the annual congress of the Swiss-based International Association of Scientific Tourism Experts (AIEST).[50] Planina not only became a member of AIEST (as did five other Yugoslav experts in 1961, among them Dušan Marić, the secretary-general of the TSJ), but also participated in the AIEST's annual meetings in Greece (1961) and Yugoslavia (1962). The latter was AIEST's first meeting in a socialist country, and it gave Yugoslavs direct access to the tourism management expertise of their European colleagues. According to the UNESCO representative, the 1962 congress covered many questions about tourism development, management, and planning at "the request of the Yugoslav guests who wanted to benefit from the experiences and knowledge of the Congress's participants."[51] Echoing this demand, an OEEC administrator said in 1960, referring to training programmes, that Yugoslavia "needed experience, skills, and facilities" and the "training to be provided . . . is designed to meet these deficiencies, thereby enabling Yugoslavia to further the development of its tourist industry and, at the same time, [increase] its ability to earn needed foreign exchange."[52] Moreover, these training programmes were meant to enable Yugoslavs to lead professional workshops – and thus train their compatriots – and to consolidate Yugoslavia's foreign market share. In keeping with these goals, the Yugoslav trainees told French journalists in 1960 that "We have found here [in France] many ideas that will allow us to equip our hotels and restaurants and to make a modern use of them that will please our Western European clients, whom we desire to see in greater numbers. . . . Our country had first to build its heavy industry. . . . The moment has come to care about tourism."[53]

The growth in foreign visitor numbers and foreign currency income in the 1950s may make their declaration seem a bit behind the times. Yugoslavia had already shown its robust commitment to the development of tourism by the time they were interviewed, and had laid the groundwork for mass consumption. On the other hand, look at the exponential growth of tourism in Yugoslavia in the 1960s, the freedom of movement Yugoslavs started to take advantage of, and the manifold impact of tourism on the culture, demographics, politics, and economics of the country, and the trainees' declaration seems more of a premonition. Like most countries in the world, Yugoslavia had just begun to care seriously about tourism.

The job awaiting Yugoslavia was monumental, and here the trainees played an even more important role, acting as economic and cultural mediators. Other (often more senior) trainees also took part in exchanges. In the spring of 1962, Stanko Marovt (a Putnik employee between the wars, Putnik deputy director in 1947–1953, and later an AIEST member) was in France for three months on an OECD grant for further training in regional tourism planning. In 1965, Ljubomir Drndić, the former director of the New York Yugoslav Information Office (1950–1953), requested to go to France for two months to train in tourism information services and propaganda as a UN fellow. Drndić was not wet behind the ears, having worked 18 years for the Yugoslav Ministry of Foreign Affairs and then as director of the Federal Committee for Tourism Information Services. Echoing the ETC's agenda, in his publications Drndić championed a tourist-friendly world with fewer bureaucratic obstacles.[54]

Given the decentralised nature of the Yugoslav tourist associations, it is difficult to gauge how many were sent to train in Western Europe, but their number grew, and they came not only from governmental agencies, but from various sectors of the Yugoslav economy. In 1965, the publishing house Turistička Štampa announced it was now the main publisher for all tourist publications, explaining that their photo service was modernised "according to the [standards of the] Kodak laboratory in Paris, where its experts were trained."[55] These networks, I would argue, contributed to the circulation of know-how, and impacted the field of tourism marketing, tourism planning, and the tourism industry as a whole. Hence one group of trainees in 1960–61 stayed first in Italy and then France, and planned to travel on to the UK and Iceland to study tourism advertising.[56] Another Yugoslav group in Geneva in 1961 was greeted by Robert Lonati, the IUOTO secretary-general, who agreed to put them in contact with the Alliance Internationale du Tourisme, the Swiss Touring Club, and the Yugoslav diplomat Vladimir Velebit.[57] Velebit was then the secretary-general of the United Nations Economic Commission for Europe (UNECE), tasked in his words, "to favour international economic cooperation." "For seven years, I led this institution, which . . . to some extent played the role of an observation outpost for its members: observation of Eastern European countries for Westerners, observation of economic and technological evolution for socialist countries. East–West contacts were an essential part of the UNECE's work."[58] Velebit's role was mainly diplomatic, but he understood, as did most of the Yugoslav economic and political leadership, that Yugoslavia had "to play the tourism card."[59]

While Yugoslavia's participation in the ETC was an early example of the country's attempts to insert itself into the Euro-Atlantic tourism network, the Soviet

Union also began opening Eastern Europe to international tourism and sought to join international tourism organisations. Soon after the 1955 Geneva Summit, the Soviet Union, Poland, and Romania joined the IUOTO, and the first two asked to become members of the ETC since their IUOTO memberships made this possible. In considering their request, the ETC's members initially unanimously accepted the Polish National Tourism Office, but postponed and finally rejected Poland's participation in the joint publicity campaigns (all while the Soviet case was pending).[60] On the one hand, Haulot and the ETC viewed the Soviet and Polish requests to be positive moves towards freedom of movement for tourists; on the other hand, they also shared concerns about the effect Poland's membership would have on the joint publicity campaign. At a broader level, its membership would have also created a precedent, opening the ETC's door to other Soviet bloc members. Reviewing the issue in a letter to the IUOTO president in January 1956, Haulot stressed this ambivalence:

> There is no doubt that the USSR's affiliation with the IUOTO was at that moment [October 1955] definitive evidence of Russians' wish to re-establish tourist contacts with the rest of Europe in particular. . . . I myself believe that membership of the IUOTO implies not only respect of the statutes, but more importantly the willingness to follow their spirit. The Union's doctrine is clear and unequivocal. Its action aims to make exchanges between peoples through tourism as easy as possible. . . . Poland's membership of our regional commission may raise considerable difficulties for our future action, especially for our collective propaganda in the US. . . . We should not forget that if Poland is the first to join, it will not be the last![61]

Subsequently, in early 1956, Haulot declared that Poland's membership could only be detrimental to the ETC, and suggested the creation of a regional commission composed of Eastern European countries, but the USSR rejected this idea. Facing the loss of its affiliation with the OEEC if it accepted Eastern European countries, the ETC concluded in the autumn of 1956 that it would be better to create two new organisations: an enlarged European regional commission under the aegis of the IUOTO; and a separate organisation for joint publicity campaigns by the ETC's original members.[62] Given the name recognition of "European Travel Commission," it was suggested that the joint publicity campaigns retain it, while the "new" regional commission for Europe should adopt a new name.[63] The result was the Regional Commission for European Travel (RCET), which was made up of national tourism boards from both Western and Eastern Europe, and worked on the same issues as the ETC, but with one significant difference: it did not pursue joint tourism publicity in the American market. That remained the ETC's flagship activity.

Conclusion

The ETC plainly considered Poland and the Soviet Union a liability, but Yugoslavia appeared not to have generated the same concern. On the contrary, the addition of

Yugoslavia had been heralded as a positive sign of European unity. The Yugoslav point of view was likely different, but the lack of evidence precludes a definitive conclusion on this point. Research does indicate that Yugoslavs were generally eager to work with Western partners. As for the norms that regulated the industry, Yugoslav tourism and foreign-trade experts quickly grasped the advantages of opening the country to American and European tourists. Yugoslavs sought help from American and other Western advertising agencies to promote their country, just like any Western enterprise. Yugoslavia's national branding abroad was achieved thanks to a mix of transnational processes, American advertising expertise, and the rise of tourism experts. True, local sensitivities and experience also shaped Yugoslavian practices and tourism norms, but this was downplayed in the ETC's adverts and the many guidebooks produced abroad, albeit with Yugoslav collaboration. Across Europe, conceiving, managing, and promoting tourism was acknowledged to require international collaboration, which inevitably could not be limited to the national context.

The professional training programmes for Yugoslav tourism experts indicated the level of cooperation needed. European expertise was well regarded, and the number of Yugoslavs who trained in Western Europe was large, but the American tourism management model seems to have retained its attraction for its efficiency, market orientation, and professionalism. While the Cold War is routinely, and often correctly, portrayed as a period that isolated people, it might be misleading to put too much stress on the geopolitics: the materiality of tourism development must not be overlooked, as interactions between people (and ideological systems) were also defined by market or economic considerations, tradition, technological advances, and interpersonal relations. After all, it was contacts between individuals and international organisations that may have resulted in the collaborations that transcended Cold War rivalry – even if they resulted from it in the first place.[64] In his research on the Council for Mutual Economic Assistance (Comecon), Simon Godard argues, for instance, that the Comecon and international organisations were "privileged microcosms" where networks formed and new ways of considering the national context within a global context developed.[65] Similarly, the ETC and the OEEC were international organisations that not only served a specific function for Yugoslavia by enabling it to use tourism as a formidable tool for cultural diplomacy, but also permitted the creation of networks that brought together governments, tourism associations, and tourism and advertising experts to produce the norms and values in the field of international tourism. These changes occurred in a decade (the early 1950s to the early 1960s) during which Yugoslavia witnessed a boom in foreign tourism, and transformed itself into a relatively well-established world tourist destination.

Notes

1   Until recently, the scholarship on this complex institution has been mostly limited to its role in the post-war reconstruction of Europe under the Marshall Plan. For recent, innovative research, see the work of the OECD History Project (http://oecdhistoryproject.

net), and, for a broader historical perspective, the essays in Richard T. Griffiths, ed., *Explorations in OEEC History* (Paris: OECD, 1997). For the history of the ETC, see Frank Schipper, Igor Tchoukarine, and Sune Bechmann Pedersen, *The History of the ETC (1948–2018)* (Brussels: European Travel Commission, 2018).

2 Report on the European tourism industry and the reconstruction of Europe 1948–1951 to the IUOTO by Ernest W. Wimble, July 1948, TOU (49) 7, OECD Library and Archives, Paris, France.

3 Letter from Henry Ingrand to Robert Schuman, 18 August 1950, Box 366, DE/CE/OECE Tourism Committee (1948–58), Diplomatic Archives Centre of the French Ministry of Foreign and European Affairs, La Courneuve, France. The cost of tourism promotions for the American market had prompted several European governments to work together in the interwar period.

4 Note on ETC's goals. Letter from the French tourism commissariat to the Minister of foreign affairs, 5 May 1948, Box 366, DE/CE/OECE Tourism Committee (1948–58), Diplomatic Archives Centre of the French Ministry of Foreign and European Affairs.

5 Correspondence between the secretary-general and the chairman of the ETC, 14 December 1951, C (51) 386 Relations between the OEEC and the ETC, Ministry of Foreign Affairs, GS 73 C 41 21 a VII), Danish National Archives, Copenhagen, Denmark. I wish to thank Sune Bechmann Pedersen for bringing this document to my attention.

6 Walter L. Hixson, *Parting the Curtain: Propaganda, Culture, and the Cold War, 1945–1961* (New York: St Martin's Press, 1997).

7 ETC Meeting, 29 March 1949, UIOTO – 295, United Nations World Tourism Organization Archives (hereafter UNWTO Archives), Madrid, Spain. Archival research in the IUOTO archives in Madrid was conducted by Frank Schipper, with whom Sune Bechmann Pedersen and I co-authored a book on the history of the ETC (see n. 1).

8 Letter from the president of the working group on tourism to the OECE executive committee, 8 December 1948, TOU (48) 1, OECD Library and Archives.

9 Council's recommendations on entry visa suppressions for the nationals of participating countries, 3 April 1950, p. 1, Council (C) (50) 75, OECD Library and Archives.

10 Arthur Haulot's report to the Committee for Tourism, November 1954, p. 7, TOU (54) 14, OECD Library and Archives.

11 New measures taken since June 1954, 15 November 1954, TOU (54) 12, OECD Library and Archives.

12 Arthur Haulot's report. November 1954, pp. 49–50, TOU (54) 14, OECD Library and Archives.

13 Records of the Department of State relating to Internal Affairs, Yugoslavia 1950–1954, Central File: Decimal File 868.181, US National Archives. *Archives Unbound*. Web. 2 April 2017.

14 Patrick Hyder Patterson, *Bought & Sold: Living and Losing the Good Life in Socialist Yugoslavia* (Ithaca: Cornell University Press, 2011), 4, 54–55, 83–84.

15 Patterson, *Bought & Sold*, 84.

16 Letter from TSJ general secretary Ljubo Babić to the Yugoslav Ministry of Foreign Affairs, 8 July 1954, Yugoslavia – 1954, file 47/12, Archive of the Ministry of Foreign Affairs of the Republic of Serbia, Belgrade, Serbia.

17 *XXIXe Congrès FIAV Dubrovnik 1955: Excursions à travers la Yougoslavie* (Belgrade: Putnik, 1955).

18 "Living Better: Americans Turn Travel Business into Third Largest Industry," *Tide, Newsmagazine of Sales & Advertising* (March 28, 1953): 56–58; Birger Nordholm's activity report in the US, June 10 1953, TOU (53) 1, OECD Library and Archives.

19 Purposeful tourism – holidays where tourists learn about politics and culture and serve as unofficial representatives of their country abroad – is a form of cultural diplomacy. On the ETC and purposeful tourism, see Christopher Endy, *Cold War Holidays: American Tourism in France* (Chapel Hill: University of North Carolina, 2004), 114–15.

20 Letter from the minister of Public Works and Transports to the minister of Finance, 1st office, 1945, F14–13714 (1946–48) Cabinet Moch, French National Archives, Pierrefitte-sur-Seine, France.
21 ETC Report on the 1952 joint publicity campaign, 9 December 1952, TOU (52) 7, OECD Library and Archives. The guidebook sold 50,000 copies on the American market in less than a year.
22 *Travel Key to Europe* (New York: This Week Magazine, 1952), 9.
23 Ibid., 271–72.
24 Patrick Hyder Patterson, "Dangerous Liaisons: Soviet-Bloc Tourists and the Temptations of the Yugoslav Good Life in the 1960s and 1970s," in *The Business of Tourism: Place, Faith, and History*, ed. Philip Scranton and Janet F. Davidson, 203–4 (Philadelphia: University of Pennsylvania Press, 2007), 186–212.
25 *Yugoslavia, New Land in the Making* (New York: Yugoslav Information Center, 1951).
26 Minutes of the 9th session, OEEC Tourism Committee, 31 December 1951, TOU/M(51) 3, Annex 1, Possible maintenance in 1952 of the joint publicity campaign in the United States, pp. 18–19, Ministry of Foreign Affairs, GS 73 C 41 21a VII, Danish National Archives.
27 Financial report for the 1956 advertisement campaign, p. 3, TOU (57) 9, OECD Library and Archives.
28 Advertisement campaign for 1962, 25 May 1961, TFD/TOU/166, OECD Library and Archives.
29 Ad hoc working party on joint action in connection with tourist publicity, 12 June 1962, TOU/TFD/179, OECD Library and Archives.
30 Groupe de travail ad hoc sur l'action collective dans le domaine de la propagande collective, 23 November 1962, TFD/TOU/187, OECD Library and Archives.
31 *Holiday* 18, no. 6 (December 1955): 76; *Holiday* 20, no. 5 (November 1956): 164.
32 David Stephens, "Yugoslavia Today," *Travel* 100, no. 1 (July 1953): 21, 23.
33 Collective advertisement campaign in the US from 1 January 1 to 30 April 1954, p. 13, ETC Report, TOU (54) 7, OECD Library and Archives.
34 Collective advertisement campaign in the US in 1955, 2 April 1956, ETC report, TOU (56) 8, OECD Library and Archives.
35 Collective advertisement campaign in the US in 1956, 22 May 1957, Annex, p. 30, TOU (57) 10, OECD Library and Archives.
36 Collective advertisement campaign in the US in 1956, 20 April 1959, TOU (59) 6, OECD Library and Archives.
37 Échanges touristiques Est-Ouest, 18 January 1968, TFD/TOU/277, OECD Library and Archives.
38 Endy, *Cold War Holidays*, 154.
39 Victoria de Grazia, *Irresistible Empire: America's Advance through 20th-Century Europe* (Cambridge, MA: Belknap Press, 2005), 239.
40 Patterson, *Bought & Sold*, 4, 54–55, 83–84, 89.
41 "Konferencija za štampu W. Clydea šefa američke trgovinske misije," *Vjesnik*, August 21, 1958, Fred Wittner papers, mss. 139AF, Box 2, Folder 2, Wisconsin Historical Society Archives, Madison, United States.
42 US Trade Mission's Report on Tourism Possibilities for Yugoslavia, 6 November 1958, p. 1, 9–10. USCOMM-D.C. 33246.
43 *The 1958 United States Trade Mission to Yugoslavia*, August 18–September 26, 1958, November 7, 1958, 2, USCOMM-D.C. 33247.
44 "Kraj koji oduševljava. Postoje mogućnosti da se znatno poveća posjeta američkih turista," *Pobeda*, September 14, 1958, Fred Wittner papers, mss, 139AF, Box 2, Folder 2, Wisconsin Historical Society Archives.
45 US Trade Mission's Report on Tourism Possibilities for Yugoslavia, 6 November 1958, p. 9, USCOMM-D.C. 33246. Gordon had made these suggestions in an interview in the newspaper *Turističke novine*, 4 October 1958.

46 Fred Wittner papers, mss. 139AF, Wisconsin Historical Society Archives.

47 Gordon and Wittner, "Dva Američka Mišljenja o oglašavanju u Jugoslaviji," *Pregled,* December 1958, 30, Fred Wittner papers, mss. 139AF, Box 2, Folder 2, Wisconsin Historical Society Archives.

48 Akira Iriye, *Global Community: The Role of International Organizations in the Making of the Contemporary World* (Berkeley & Los Angeles: University of California Press, 2002); Sari Autio-Sarasmo, "Khrushchev and the Challenge of Technological Progress," in *Khrushchev in the Kremlin: Policy and Government in the Soviet Union, 1953–1964,* ed. Jeremy Smith and Melanie Ilic, 133–49 (London: Routledge, 2011); Simo Mikkonen and Pia Koivunen, eds., *Beyond the Divide: Entangled Histories of Cold War Europe* (New York: Berghahn, 2015).

49 Blagoje Bogavac, "Yugoslavia and International Technical Cooperation," *Review of International Affairs* 19, no. 430 (March 5, 1968): 26. See also Bogavac, *Učešće Jugoslavije u međunarodnoj tehničkoj saradnji* (Belgrade: Savezni zavod za međunarodnu tehničku saradnju, 1968).

50 Participant Biographical Data: Janez Planina, Tourisme, Service de l'action touristique, 19790203/17 Coopération technique 1964–1970, Yougoslavie, French National Archives.

51 Rapport de A. Khoshkish, Mission au 13ᵉᵐᵉ Congrès de l'Association Internationale d'experts scientifiques du tourisme (AIEST), tenu du 10 au 15 septembre 1962 en Yougoslavie, B. 244/AIEST, UNESCO Archives, Paris, France.

52 S.J. Joyce, OEEC Third Country training section, Kitchen management and operations, 18 August 1960, p. 3, Tourisme, Service de l'action touristique, 19790203/17 Coopération technique 1960–1974, Yougoslavie, French National Archives.

53 "Envoyés par leur gouvernement, quatre spécialistes yougoslaves du tourisme visitent la Haute-Savoie," *Annecy,* November 25, 1960, 1.

54 Ljubo Drndić, "Yugoslavia – An Open Social Community," *Review of International Affairs* 16, no. 363 (1965): 16.

55 Informacija o važnijim aktivnostima i proizvodnji u Turističkoj štampi, 24 May 1965, Fund 580, Federal Committee for Tourism, Box 37, Archives of Yugoslavia, Belgrade, Serbia.

56 Note du Commissariat général au tourisme, coopération technique, groupe n° 830 – Yougoslavie, 2 December 1960, Tourisme, Service de l'action touristique, 19790203/17 Coopération technique 1960–1974, Yougoslavie, French National Archives.

57 Lettre de Robert Lonati à M.G. Le Brec du Commissariat général au tourisme, 8 February 1961, Tourisme, Service de l'action touristique, 19790203/17 Coopération technique 1960–1974, Yougoslavie, French National Archives.

58 Jean-François Berger, ed., *Dans l'ombre de Tito: Entretiens avec le général Vladimir Velebit* (Geneva: Slatkine, 2000), 139.

59 Ibid., 124–25.

60 Provisional minutes of the ETC meeting in New Delhi submitted by Haulot, 19 October 1955, IUOTO – 296, UNWTO Archives.

61 Letter from Haulot to the IUOTO, confidential, 20 January 1956, IUOTO – 296, UNWTO Archives. See also Haulot's correspondence with Karchenko and Ankudinov from the Soviet travel agency Inturist.

62 ETC Meeting, 23–25 May 1956, IUOTO – 295, UNWTO Archives.

63 Mémorandum relatif aux notes de la Pologne et de l'URSS, 10 September 1956, IUOTO – 295, UNWTO Archives.

64 See, for instance. Sandrine Kott, "Par-delà la guerre froide: Les organisations internationales et les circulations Est-Ouest (1947–1973)," *Vingtième Siècle* 1, no. 109 (2011): 142–54.

65 Simon Godard, "Une seule façon d'être communiste? L'internationalisme dans les parcours biographiques au Conseil d'aide économique mutuelle," *Critique internationale* 66, no. 1 (2015): 69–83.

References

Autio-Sarasmo, Sari. "Khrushchev and the Challenge of Technological Progress." In *Khrushchev in the Kremlin: Policy and Government in the Soviet Union, 1953–1964*, edited by Jeremy Smith and Melanie Ilic, 133–49. London: Routledge, 2011.

Berger, Jean-François, ed. *Dans l'ombre de Tito: Entretiens avec le général Vladimir Velebit*. Geneva: Slatkine, 2000.

Bogavac, Blagoje. *Učešće Jugoslavije u međunarodnoj tehničkoj saradnji*. Belgrade: Savezni zavod za međunarodnu tehničku saradnju, 1968.

———. "Yugoslavia and International Technical Cooperation." *Review of International Affairs* 19, no. 430 (1968): 24–27.

de Grazia, Victoria. *Irresistible Empire: America's Advance through 20th-Century Europe*. Cambridge, MA: Belknap Press, 2005.

Drndić, Ljubo. "Yugoslavia – An Open Social Community." *Review of International Affairs* 16, no. 363 (1965): 15–16.

Endy, Christopher. *Cold War Holidays: American Tourism in France*. Chapel Hill: University of North Carolina, 2004.

Godard, Simon. "Une seule façon d'être communiste? L'internationalisme dans les parcours biographiques au Conseil d'aide économique mutuelle." *Critique Internationale* 66, no. 1 (2015): 69–83.

Griffiths, Richard T., ed. *Explorations in OEEC History*. Paris: OECD, 1997.

Hixson, Walter L. *Parting the Curtain: Propaganda, Culture, and the Cold War, 1945–1961*. New York: St Martin's Press, 1997.

Iriye, Akira. *Global Community: The Role of International Organizations in the Making of the Contemporary World*. Berkeley & Los Angeles: University of California Press, 2002.

Kott, Sandrine. "Par-delà la guerre froide: Les organisations internationales et les circulations Est-Ouest (1947–1973)." *Vingtième Siècle* 1, no. 109 (2011): 142–54.

Mikkonen, Simo, and Pia Koivunen, eds. *Beyond the Divide: Entangled Histories of Cold War Europe*. New York: Berghahn, 2015.

Patterson, Patrick Hyder. "Dangerous Liaisons: Soviet-Bloc Tourists and the Temptations of the Yugoslav Good Life in the 1960s and 1970s." In *The Business of Tourism: Place, Faith, and History*, edited by Philip Scranton and Janet F. Davidson, 186–212. Philadelphia: University of Pennsylvania Press, 2007.

———. *Bought & Sold: Living and Losing the Good Life in Socialist Yugoslavia*. Ithaca: Cornell University Press, 2011.

Schipper, Frank, Igor Tchoukarine, and Sune Bechmann Pedersen. *The History of the ETC (1948–2018)*. Brussels: European Travel Commission, 2018.

Stephens, David. "Yugoslavia Today." *Travel* 100, no. 1 (July 1953).

Travel Key to Europe. New York: This Week Magazine, 1952.

XXIXᵉ Congrès FIAV Dubrovnik 1955: Excursions à travers la Yougoslavie. Belgrade: Putnik, 1955.

9 Making Iron Curtain overflights legal

Soviet–Scandinavian aviation negotiations in the early Cold War

Karl Lorentz Kleve

Aviation technology took tremendous leaps forward during the Second World War. In the 1950s, the new advances led to a rapid expansion of civil aviation. The technology allowed for the opening of commercially viable transcontinental routes. In Europe, Scandinavian Airlines System (SAS) was at the forefront of developments, taking advantage of its local experience of Arctic airspace as it pioneered routes to North America and later to the Far East.

At the same time, the 1950s was also a time of military build-up on both sides of the Iron Curtain. The only directly shared border in Europe between the Soviet Union and a NATO member was at the northernmost point of the Scandinavian Peninsula. Northern Norway and the Soviet-controlled Kola Peninsula thus underwent radical transformations from backward fishing and peasant societies to modern industrial regions, home to massive military installations. The opposing alliances' airports, missile sites, and fleet facilities were sometimes just a few miles apart. Both sides spied intensely on each other, and by the end of the 1950s both NATO and the Soviet Union undertook regular intelligence flights along each other's borders, attempting to penetrate the veil of secrecy. In 1960, the area famously hit the headlines when the Soviets shot down an American U-2 spy plane on its way from Peshawar in Pakistan to Bodø in Norway.

In this atmosphere of militarisation and political distrust, however, Norway and the Soviet Union also negotiated a legal framework for commercial overflights and landing rights in each other's territory. In order to develop regular routes to the Far East, being able to overfly Soviet airspace was of great importance to the Western European carriers. In turn, for the Soviet Union, being able to overfly the Scandinavian Peninsula was essential to establishing intercontinental routes to North America and Cuba. Hence, both states had a strong interest in reaching an agreement on aviation and flying rights.

This chapter is a study of the diplomatic process leading up to the successful negotiation of an aviation agreement between Norway, Sweden, and Denmark and the Soviet Union in March 1956. It analyses how the agreement came about and the consequences it had for aviation between the blocs during the Cold War. The chapter also considers the interwoven role of the three Scandinavian states and their pioneering multinational flag carrier, SAS, in the 1956 negotiations and subsequent talks. Although the agreement was shaped by Cold War politics it in fact remains in force to this day. The final part of the chapter therefore discusses the lasting implications of the agreement for contemporary aviation over Russian airspace.

The birth of an international aviation regime

The end of the Second World War set the stage for concerted attempts to establish a new world order.[1] With the end of the war in sight, 54 nations, including all the Western allies, signed the Chicago Convention in December 1944. The Convention established the basic principles governing commercial aviation and founded the International Civil Aviation Organization (ICAO), which in 1947 became a UN specialised agency.[2] Most of the delegations at the Chicago Convention included airline representatives as advisers. Thirty-four of these airlines also met separately to establish a non-governmental airline association. At a second airline conference at Cuba in April 1945, the International Air Transport Association (IATA) was thus founded.

Even though optimism was high regarding the possibility of creating a truly international aviation regime, the primary result of the Chicago Convention was the confirmation of the right of states to control their own airspace, as well as their right to government involvement in commercial aviation. This meant that practical matters regarding the opening and maintenance of air routes needed to be settled in bilateral negotiations between states. The so-called "Five Freedoms" agreed by the participants concerned each state's right to award airlines from other countries the freedom of transit, landing, taking on cargo and passengers, of unloading cargo and passengers, and of bringing passengers or cargo to and from third countries. These freedoms were to be confirmed through bilateral negotiations.[3] Throughout most of the Cold War, aviation remained one of the most state-regulated businesses in the world.[4] Each nation acted as fully sovereign within its own borders, and thus could grant the freedoms agreed upon in Chicago as they saw fit.

Aviation was certainly not the only area where many nations had what the sociologist and negotiation theorist Anselm Strauss called "overriding common stakes."[5] Between Western countries where the level of trust was quite high, international agreements on a wide range of areas were made in short time in the 1940s, bilateral aviation agreements among them. The negotiations for aviation agreements are good examples, though, of Anselm's "overriding common stakes" eventually making agreements possible despite a certain level of distrust and opposition. Under Stalin, the fear of spies ruled out Western collaboration in civil aviation, even though the Soviet Union did recognise its potential. Soviet distrust was matched by a deep suspicion in the West of allowing foreign operators access to national airspace. When the Czechoslovak airline CSA applied for an air route between Prague and Oslo in 1952, Wilhelm Evang, Head of Norwegian Military Intelligence, told the Foreign Ministry that he could see no reason to grant this application, as the only people needing an air route from Prague to Oslo would be spies. The application for this particular route was subsequently denied by the Norwegian Foreign Ministry.[6]

The first major bilateral negotiations in the world took place between the two leading airpowers of the time, the UK and the US. The resulting Bermuda Agreement of 1946 established a precedent for other countries. Subsequent years saw a flurry of activity as states and airlines hurried to negotiate agreements and start flying. Aviation was the future and no country wanted to be left behind. After or parallel to successful negotiations, the airlines involved usually conducted their own negotiations to settle more practical matters such as route schedules, access to fuel and maintenance, the organisation of ticket sales, and so on. With its vast size and strategic location, access to Soviet airspace was a prize vied for by most Western airlines. However, mounting Cold War antagonism rendered negotiations difficult.

Past Norwegian–Soviet negotiations

The idea that Norway might play a key role in a transcontinental network of air routes predated the Second World War. In 1938, three years after being awarded the first public air route concession, Det Norske Luftfartsselskap (DNL, the Norwegian Aviation Company) struck a deal with British Imperial Airways and Irish Rianta for a transatlantic route to the US from the brand-new Sola Airport outside Stavanger, via Shannon in Ireland. DNL envisioned future possibilities for a transglobal route linking the US and the Soviet Union via Norway.[7] DNL's international agreements prompted Danish, Finnish, and Swedish airlines to approach DNL for talks about transatlantic cooperation. At a meeting held in Berlin in 1939, the airlines agreed that a joint Nordic consortium would be a stronger player in negotiations with the US and the Soviet Union.[8] The outbreak of the war soon put an end to this idea, though.

During the war, DNL's assets were seized by the occupying Germans, while its pilots and administrators escaped and joined the Norwegian Armed Forces abroad. Towards the end of the war, the Norwegian government in exile in London established a new government subsidiary, the Norwegian Aviation Board, to prepare a restart of civil aviation after the war.[9] The government in exile also informed the Soviet authorities in January 1944 that it would like to establish aviation connections once the war was over.[10] The Soviets welcomed the idea that same month. In fact, Sweden had already made a similar request, but the Chief of the Nordic Division in the Soviet Foreign Ministry, Mr Sergeev, and the Norwegian ambassador in Moscow, Rolf Andvord, had agreed that connections with Norway were much more important. After all, Norway and the Soviet Union were allies in the war against Germany, and Scandinavian cooperation in the air at this stage was not as obvious as it had been before the war, or as it would later be.[11]

By 1945, however, the atmosphere in the Soviet Union had changed. Answering a request for negotiations on an air route between Oslo and Moscow in the autumn of 1945, Assistant Foreign Minister Dekanozov stated categorically that the Soviet Union did not give air concessions to foreign countries, and that it did not accept foreign air routes to cross its borders. For example, existing British and US air routes to the Soviet Union ended in Teheran. Swedish diplomats seem to have told the Norwegian ambassador that they were expecting Swedish routes to Moscow to commence any time soon. Dekanozov, however, denied this, and the ambassador reported home to the Norwegian Foreign Ministry that the case of routes to Moscow was a delicate matter.[12]

In March 1946, DNL was re-established and sought to set up a weekly courier route to Moscow. Joachim G. Urby, a DNL pilot and general manager, told the Norwegian Foreign Ministry that the Soviet ambassador in Oslo was positive.[13] Accordingly, a delegation from DNL, the Foreign Ministry, and the Ministry of Transportation went to Moscow to start negotiations with the Soviet Union on 4 May 1946. Even though the routes to the Soviet Union were less important to DNL than the routes to Scandinavia and Western Europe, it was still a priority to lay foundations for the future.[14] The Soviets were willing to negotiate, but flatly rejected the possibility of foreign aeroplanes in Soviet airspace, except for newly conquered Klaipeda, the former German city of Memel. The Soviets offered Norway a route to Klaipeda from which Soviet planes and staff would take over on the final leg to Moscow. Norway probably would have accepted this, had the Soviet Union not demanded the right to routes all the way to Oslo in return.[15]

This meant that the Soviets were not going to adhere to the principle of reciprocity, which had just been established by the Chicago Convention and the Bermuda Agreement as the basis for international negotiations. By this point, several other Western countries had also started negotiations with the Soviet Union, and it appears that they all encountered the same demands: the Soviets wanted access all the way to the Western capital, while the Western side was not allowed to operate beyond Klaipeda. The Soviets seemed to fear that civilian aeroplanes would be used for spying, and after two months the negotiations collapsed. The only Western country to make inroads was Sweden. This deal resulted not from an official bilateral agreement, but from a company-based agreement between the Swedish airline ABA and the Soviet airline Aeroflot. It entailed a route from Stockholm to Moscow via Helsinki, with Swedish planes flying Stockholm–Helsinki and Soviet planes flying Moscow–Helsinki.[16]

No other air agreements were reached for the remainder of Stalin's reign, even though the Soviet ambassador in Oslo twice proposed a restart of negotiations. In 1947 and in 1948 he suggested to Foreign Minister Molotov that an aviation agreement might be a part of a diplomatic thrust to counter the growing British and American influence in Norway.[17] Yet nothing came of the initiative, probably owing to the growing mistrust and fear of spying on either side of the Cold War divide.

New winds from the East

With Stalin's death in 1953 and Khrushchev's rise to power in 1955, East–West relations relaxed considerably. Norwegian Prime Minister Einar Gerhardsen, his wife Werna and Trade Minister Arne Skaug visited Moscow in October 1955. The trip caused an internal dispute within the ruling Labour Party, and it was strongly opposed by Foreign Minister Lange and other leading Labour politicians, who feared that Khrushchev would try to convince Gerhardsen to leave NATO and collaborate more closely with the Soviet Union. The British ambassador in Oslo, Sir Peter Scarlett, worried that the Soviet Union was wooing Norway by offering it a special treatment – a sentiment he voiced in a letter to Foreign Minister Selwyn Lloyd immediately prior to Gerhardsen's trip.[18] Scarlett recommended that Gerhardsen be invited to Britain afterwards, to be "educated." In a recent analysis of the visit, however, the historian Stian Bones concludes that Gerhardsen was only trying to improve the bilateral relations now that Khrushchev's leadership made rapprochement between Norway and the Soviet Union possible.[19]

The thaw put air routes across the Iron Curtain back on the diplomatic agenda, although the question was not officially mentioned during the visit. A letter from Soviet Prime Minister Bulganin of 19 March 1957 to Gerhardsen – made public in the press in both Norway and the Soviet Union a week later – declared, however, that the aviation agreement was one of the tangible results of the trip.[20] Indeed, just one month after the visit of the Norwegian Prime Minister, SAS CEO Henning Throne-Holst also travelled to Moscow, in order to discuss an extension of the Swedish 1946 ABA–Aeroflot agreement with the Soviet airline. The journey was evidently planned a couple of weeks beforehand, following an invitation from Aeroflot. This was no secret, although SAS does not seem to have informed all its owner companies prior to Throne-Holst's Moscow visit: there is no evidence in the archives that the CEO wrote to the Norwegian Foreign or Transport Ministry, nor do the minutes of the SAS board mention the upcoming trip. Director Boye of the Norwegian regional SAS office informed the Foreign Ministry by phone, and only once Throne-Holst was already in Moscow. He also said that the Soviet Union still seemed to be reluctant to allow aeroplanes and personnel from NATO countries into Soviet airspace.[21]

Throne-Holst's visit to Moscow was probably driven by two reasons. The first was the Soviet Union's renewed interest in Western connections, which led to aviation agreements struck with Finland on 19 October 1955, followed by agreements with Austria on 9 November and Yugoslavia by the end of that month.[22] Soviet newspapers hailed the agreement with Finland as the very first of its kind, ushering a new era in air travel.[23] The second was the change in Scandinavian aviation between 1946 and 1955. On 1 August 1946, DNL, ABA, and the Danish airline DLL had signed the SAS Agreement that established a joint company named SAS Overseas, designed to handle transatlantic traffic between Norway, Sweden, Denmark, and North America. On 8 February 1951, the three national airlines became purely holding companies, and SAS was made responsible for all international and domestic traffic. Any renewal of the ABA–Aeroflot agreement would therefore require the inclusion of Denmark and Norway.

Actually, there had been a renewal of sorts of the Swedish ABA–Aeroflot-agreement covering SAS two years earlier, in 1954. On 1 October 1953, the Norwegian embassy in Moscow informed the Foreign Ministry that two SAS employees had travelled to Moscow for discussions with Aeroflot, to moot a joint venture route between Stockholm and Moscow via Helsinki and Leningrad. The embassy warned that bilateral agreements between the Soviet Union and

Denmark and Norway might become necessary for these discussions to proceed.[24] Nevertheless, the negotiations seem to have led to a continued company-based agreement between ABA and Aeroflot concerning the route Stockholm–Helsinki–Leningrad–Moscow. The agreement did not imply any SAS traffic over Soviet soil, as SAS was to use the route up to Helsinki only, where Aeroflot would take over as before. What was new was that both airlines could sell through tickets to Moscow or Stockholm respectively.[25]

Whether Throne-Holst had anything to do with the Swedish–Soviet amendment in 1954 is unclear. He had become a member of the SAS board in 1954 and CEO of SAS in 1955, so it is quite likely. A memo of 2 February 1956 from the Norwegian ambassador Erik Braadland in Moscow to the Foreign Ministry, mentioned that Throne-Holst had begun thinking about a new deal with the Soviet Union early in 1955, and that he thought it would be in the Soviets' interest to reach an agreement.[26]

Why would the Soviets be interested? Because they needed access to Scandinavian airspace for routes onwards to the British Isles and North America. In Braadland's view, it was not necessary to appear too eager for an agreement at any price, as it was the Soviet Union that was boxed in on its western borders. The Soviets could fly to Eastern Europe, but in order to get further west, they needed access either to West German or Scandinavian airspace. Crossing West German airspace was out of the question, since Soviet aeroplanes would first have to cross East German or Czechoslovak airspace, and West Germany recognised neither country and could therefore not engage in bilateral aviation negotiations with them.

Even so, the Soviets appeared somewhat reserved at first. Throne-Holst initially suggested a renewal of the ABA–Aeroflot deal, using only Swedish-registered SAS planes and pilots.[27] In reaction to this, there was a meeting on 28 November between the Norwegian Foreign and Defence Minister and the new head of the Norwegian Division of SAS, Nils Langhelle.[28] Those present agreed that Norway would not block a Swedish–Soviet agreement involving the Swedish-registered parts of SAS. Such an agreement, however, would give the Soviet Union access to Swedish airspace only. If the Soviets wished to fly to or over Norway, they would need an agreement with Norway.[29]

Soviet–Scandinavian air rights negotiations in the 1950s

Throne-Holst's November meeting in Moscow led to formal Swedish–Soviet negotiations in January 1956. According to Norwegian sources, the Swedish negotiators preferred an agreement that included the whole of the SAS over a purely Swedish–Soviet agreement. Towards the end of the month, the Soviet negotiators for the first time indicated that they might also be interested in an agreement with Norway and Denmark.[30] Soviet diplomats continued to tell the press that they preferred a purely Swedish–Soviet agreement, due to Norway's and Denmark's NATO membership.[31] Behind closed doors, though, they suggested to Swedish, Danish, and Norwegian diplomats that they put the Swedish–Soviet negotiations on hold for a month to await possible Norwegian and Danish overtures.[32]

At the same time, the two sides were apparently busy planting stories in the Norwegian newspaper *Verdens Gang* (VG) in order to steer the negotiations. One article mentioned how previous agreements with the US had restrictions on Swedish pilots, probably due to Sweden's neutral stance in the Second World War and after. What was left unsaid but had been implied was that some restrictions could be acceptable in a deal with the Soviet Union. Moreover, the article warned that if an agreement was not reached quickly, the Finnish airline Aero might "steal" the traffic, since Finland had already negotiated an agreement.[33] Perhaps it had been the Soviet Union that had attempted to peddle such notions in parts of the Norwegian press. Another article in *VG* toned down the need for an agreement, though, and accused the Soviet Union of trying to sow discord.[34]

In fact, the secret Soviet invitation to negotiate triggered a fast and positive response from all three Scandinavian countries. As early as 10 February 1956, the Norwegian embassy in Paris was instructed to brief its NATO partners there on the planned negotiations.[35] Moreover, the three Scandinavian governments closely aligned their positions. Sweden sent Norway and Denmark detailed accounts of its earlier negotiations with the Soviet Union.[36] The sending of a Swedish note to the Soviet Union, officially asking for negotiations, was closely coordinated with Denmark. For the Soviet Foreign Ministry, Mr Gribanov made it clear to the Norwegian embassy that rumours in the press suggesting that Norwegian NATO membership was an obstacle to an agreement were false.[37]

On 24 February, the Norwegian Foreign Ministry published a press release about the forthcoming negotiations.[38] On 9 March, the Norwegian Council of Ministers was officially informed that an agreement had been reached with the Soviet Union to start negotiations on 21 March.[39] The Norwegian ambassador was authorised to sign a deal if it were done before Easter, and if Denmark and Sweden were also ready to sign at the same time.[40]

The negotiations were conducted jointly with Denmark and Sweden, and it was intended that the joint company SAS should operate on each Scandinavian country's behalf instead of the original national airlines, DNL, DDL, and ABA. In fact, by 1956 these airlines existed in name only, as the formal joint owners of SAS. The teams from Norway, Sweden, and Denmark negotiated with the Soviet Union together, but when the agreement was signed on 31 March 1956, the four parties entered separate yet identical agreements, thus keeping them strictly bilateral. Therefore, a note was added to the agreement stating that the Soviet Union accepted the suggestion (made in a special letter from the Norwegian ambassador) that DNL should use aeroplanes and personnel from SAS, be they Norwegian, Danish, or Swedish.

The agreement was extremely specific by modern standards, and stated that all matters regarding flights between the two states, such as the price of tickets, time schedules, technical running of operations, or the number of personnel stationed (four Norwegians in Soviet Union and four Soviet citizens in Norway) should be specified in particular agreements between the two airlines.[41] The employees whom DNL were allowed to station in the Soviet Union to maintain operations could be SAS staff, though SAS as a whole could only station four employees in the Soviet Union, while the Soviet Union could deploy four in each of the three Scandinavian countries. Since DNL, DDL, and ABA were now shell companies without employees this clause stretched the principle of reciprocity: the Soviet Union was allowed to deploy 12 people in Scandinavia while the Scandinavian countries, through SAS, could only station four in the Soviet Union.

On the same date as the bilateral agreements were signed, SAS also concluded a company agreement with Aeroflot. This agreement covered maintenance issues such as rules for discounted tickets, ticket systems, conditions of carriage, and so on. The agreement referred to the bilateral agreements, showing that both airlines were well informed of the negotiations before the actual signing.[42]

These speedy negotiations were followed by an equally speedy implementation process. The agreed-upon routes opened almost immediately. Soviet aeroplanes could operate on the Riga–Stockholm–Oslo, Moscow–Stockholm–Oslo, and Leningrad–Helsinki–Stockholm routes; Norwegian aeroplanes, Oslo–Stockholm–Riga, Oslo–Stockholm–Riga–Moscow, and Oslo–Stockholm–Helsinki–Leningrad. The agreement specified that the planes operating had to be owned and manned by Norwegian or Soviet personnel respectively. It also stressed that all rights and regulations for the airlines' operations should be reciprocal. This would prove an obstacle later, as it meant that if Norway wanted to upgrade its operations to larger planes than the Soviet Union wanted to operate, they would not be admitted on Soviet routes. All aeroplanes had to be comparable in size and capacity. If the Soviets used a twin-engine 44-seater aeroplane, Norway could not upgrade to a four-engine 80-seater.

That same summer, SAS began pushing the Soviet Union for permission to fly the more modern Convair planes instead of the already old-fashioned Scandias mentioned in the agreement. The Soviet Union announced that it might consider starting tests by flying its first proper jet passenger aircraft, the TU-104, on the route to Copenhagen, thereby making it possible for SAS to use the larger Convair in return. However, the Soviet Union claimed that the runway at Copenhagen's Kastrup Airport was too short for its jet. It took until the winter schedule started in October 1957 before SAS finally obtained permission. By then, the even newer and larger DC 6B had also been permitted to operate on the Moscow route in December, just as Aeroflot finally started using TU-104s on the Copenhagen route. In November 1956, the Norwegian Ministry of Transportation asked the Norwegian Foreign Ministry to lean on the Soviets to agree to an expansion of the route network to include another route – Oslo–Stockholm–Helsinki–Narva–Velikie Luki–Moscow – as an alternative route and airports were needed to circumvent weather constraints. This turned out to be difficult, though. The Soviets were not prepared to expand the number of routes and airports that fast.[43]

The final agreement did not include overflying, or transit rights to points beyond. Ambassador Braadland had initially thought that obtaining this would be the primary objective of the Soviet negotiators. However, the question only came up in the talks when the Scandinavian team raised the issue, asking whether the Soviet authorities intended to start air traffic to points beyond the ones stated in the routes list. The Scandinavians signalled they were interested in such routes via

Soviet airspace themselves, primarily to Pakistan or India, but also to Tokyo and Beijing. Aeroflot's director Marshal Zhavoronkov replied that the Soviet Union was also interested in points beyond. At this stage, however, such routes had too many problems and would better be left to future negotiations.[44] This came as a surprise for the Scandinavian negotiating teams, and it continues to be something of a mystery. By 1956, when the Soviet Union was finally ready to enter into aviation agreements, its main concern was not as ambassador Braadland had believed, transit rights through Norwegian airspace to third countries, but rather just routes between the USSR and Norway. Similar limits were put on subsequent agreements with other Western nations. And only three Soviet cities were opened to foreign routes: Moscow, Leningrad, and Riga. One could speculate why, but I would argue that the main reasons for this somewhat baffling lack of Soviet concern was the old fear of giving foreign aeroplanes access to Soviet airspace. One hallmark of aviation is the difficulty of controlling what an aeroplane actually does in the sky. The fear of aerial espionage was strong, and the Soviet Union was not prepared to let foreign aeroplanes deeper into its airspace.

Two years later, though, on 31 August 1958, the Soviet Union did sign an addition to its aviation agreement with Denmark, allowing Danish planes to cross European Soviet airspace en route to the Middle East. In return, Soviet planes were permitted to transit Denmark on their way to the UK, France, the Netherlands, and Belgium.[45] Remarkably, no such addition was agreed with Sweden or Norway.

In a comparative perspective, there were no significant differences between the Norwegian–Soviet agreement and the bilateral agreements the Soviet Union signed with other Western countries in the following months and years. The British ambassador's fears that the Soviet Union would woo Norway into closer relations were unfounded. Norway did not receive any special privileges in the air rights negotiations. Instead, it was the Soviet side that changed its position, having warmed to the idea of cooperation in aviation – as the Soviet ambassador in Oslo admitted during a dinner with the Norwegian Foreign Minister Hallvard Lange in March 1957.[46]

The agreement kept to the fiction of dealing with national operators, even if the Scandinavians had merged them in one company, SAS, in 1946. It was closely modelled on the two previous aviation agreements the Soviet Union had signed with Austria and Finland. Indeed, emulating the Austrian agreement was a stated ambition of the Scandinavian negotiators.[47] In practice, the agreement with the Scandinavian countries was unique insofar as SAS was the flag carrier of three nations simultaneously. But the style, content, and limitations mirrored most other Soviet bilateral aviation agreements.

Lasting impacts

In 1958, the recently retired head of SAS Overseas, Hjalmar Riiser-Larsen, was travelling to Japan on the airline's new route from Copenhagen to Tokyo to lecture on polar aviation. As a young man in the 1920s, Riiser-Larsen had attempted to cross the North Pole by air several times, before he finally succeeded in 1926 with Roald Amundsen and Umberto Nobile. Having a keen eye for PR, Riiser-Larsen made a short documentary film of the flight. Titled "Over Nordpolen," and thereby awakening memories of past heroic attempts to traverse the inhospitable Arctic, this journey, however, was luxurious. The food and drink was good, and the cabin even had a bed for passengers who wanted to rest. Moreover, the journey lasted only a little over 30 hours. Riiser-Larsen marvelled at the wonders of modern aviation, which had made the North Pole a perfectly navigable airspace.[48] In the beginning of the film, though, the plane is seen to take a considerable detour via Alaska rather than flying directly over the Pole, which would have required permission to cross Soviet airspace.

Operating flights to the Soviet Union was not the main interest of the Norwegian negotiators in 1956. The bilateral agreements reached by the Soviets and the Scandinavians were regarded as stepping stones towards overflight rights, even if negotiations on that particular issue were postponed in 1956, and the Siberian route had to wait another 15 years. Meanwhile, the Norwegian–Soviet aviation agreement was officially revised and supplemented in 1967, 1971, 1976, 1981, 1987, 1988, 1989, and for the last time in 1990.[49] Its core regulations remain in force, and after 1990 there have been several, albeit sporadic, negotiations that have resulted in some informal agreements on changes and revisions. The only formal revision of the agreement after 1990 was to change the country name of the Soviet Union to Russia.[50]

The main issues driving these negotiations during the Cold War were aeroplane size and permission to transit Siberia for routes to the Far East. For the Soviet Union, by contrast, transit of Norwegian airspace to Cuba was the highest priority. The 1967 revision gave the Soviet Union the right to transit Scandinavia when bound for the US, Canada, and "one point in the Americas south of the USA." In return, SAS received permission to cross Soviet airspace via Tbilisi or Tashkent towards Asia, known as the trans-Asian route.[51]

One thing worth noting about the 1967 revision was the small but very important difference between the Norwegian and the Swedish–Danish agreements. The agreement with Norway specified that the eventual route which the Soviet Union might open to America outside the US or Canada had to include a transit landing in the US or Canada. In practice, that made it impossible for the Soviet Union to use Norwegian airspace for direct flights to Cuba. In his presentation of the agreement to the Council of Ministers in February, the Norwegian Transport Minister also categorically stated that "the route cannot continue to Havana."[52] It was imperative for the US that the Soviet Union not be given easy air access to Cuba. Plainly, it could not deny Soviet air traffic via international airspace north of Norway, but to be able to cross Norwegian airspace towards Cuba would significantly shorten the route. The US Foreign Ministry asked Norway not to grant the Soviets this and Norway complied.[53] Denmark and Sweden did not yield to US pressure, so the Soviet Union could cross Danish airspace to Cuba without having to land in the US or Canada first. This was a more complicated route, though, involving either a negotiated path along the English Channel or a sharp turn north in the

North Sea to avoid British airspace. Formally, Russia is still only allowed to transit Norway to Cuba if Russian planes make a transit landing in the US.

The 1967 revision also contained a Soviet concession that allowed SAS into Siberia when a corridor for civilian traffic became available. When this happened in 1971, it was a major asset for Scandinavian aviation.[54] The opening of the Copenhagen–Tokyo route on 3 April 1971 cut flying time by a whopping thirteen hours.[55] Nevertheless, it would be another decade before wide-bodied jets such as the DC-10 or Boeing 747 were permitted to operate on this route.[56] The Siberia corridor continued to be the trickiest part of Scandinavian negotiations with the Soviets/Russians.[57] The Soviet Union was loath to open Siberia at all for security reasons. Keeping Siberia as closed as possible to foreign eyes was (and, it seems, still is) an important security consideration. The downing of the civilian Korean aeroplane KAL 007 in 1983 was testimony to the deadly risks of not following Soviet regulations to the absolute letter.

Access to Russian airspace remains essential to contemporary aviation as traffic to Asia is growing. However, Russia has continued its traditional conservative and highly regulatory stance on aviation rights. The original part of the 1956 agreement that gave SAS a monopoly on Norwegian aviation in Russian airspace remains in place. Since the agreement specified that the planes operating had to be owned and manned by Norwegian or Soviet personnel respectively, low-cost carriers such as Norwegian Airlines operating with sub-companies and personnel registered in several different countries are barred from Russian airspace. Ever since the end of the Cold War, Scandinavian governments have sought to negotiate more flexible terms with Russia. In 2017, Norwegian Airlines was finally allowed to operate a trans-Asian route over Russia, but the trans-Siberian route remains open to SAS only, as a curious remnant of Cold War politics and old monopolistic aviation regimes.

Notes

1 International agreements on commercial aviation had barely begun before the Second World War. In 1919, airlines from Norway, Denmark, Sweden, Germany, and the UK met in The Hague to found the International Air Traffic Association, a predecessor to the International Air Transport Association (IATA) established by several Western airlines in 1945. Another airline-based agreement, the Warsaw Convention of 1929 (reframed as the Montreal Agreement of 1966 and Montreal Convention of 1999), limited airline liabilities on international routes.
2 For the birth of international aviation, see Peter P. C. Haanappel, *Ratemaking in International Air Transport* (Deventer: Kluwer, 1978), 9–32; Andreas Papatheodorou, "The Impact of Civil Aviation Regimes on Leisure Travel," in *Aviation and Tourism*, ed. Anne Graham, Andreas Papatheodorou and Peter Forsyth (Farnham: Ashgate, 2008), 49–57; Arne Rosenberg, *Air Travel within Europe* (Stockholm: National Swedish Consumer Council, 1970), 26–38.
3 Rosenberg, *Air Travel*, 28.
4 Audun Tjomsland and Kjell G. Wilsberg, *Mot Alle Odds* (Oslo: Braathens SAFE, 1995), 50.
5 Anselm Strauss, *Negotiations: Varieties, Contexts, Processes and Social Order* (San Francisco: Jossey-Bass, 1978), 159.

6   Wilhelm Evang, Head of Norwegian Military Intelligence to the Norwegian foreign and defence ministers, 5 January 1953, Box 3556, Norwegian Foreign Ministry Archives (UD) 1950–59, Norwegian National Archives Riksarkivet (RA), Oslo, Norway.

7   Unsigned note marked "Oslo 7.5.1943" (probably written by de facto DNL Head of Operations in 1940–43, Leif Villars-Dahl), Box "DNL's utvikling frem til SAS – Diverse brev og dokumenter 1945–47," SAS Museum Archive, Gardermoen, Norway.

8   Ibid., 7

9   Norwegian Aviation Board to the Norwegian government in exile in London, 22 October 1945, box marked "DNL's utvikling frem til SAS – Diverse brev og dokumenter 1945–47," SAS Museum Archive. The Norwegian Aviation Board was founded in exile in London in 1943.

10 Norwegian Ministry of Defence in exile in London to the Foreign Ministry, 6 January 1944, Box 12017, UD 1940–49, RA.

11 Ambassador Andvord to the Norwegian Aviation Board, 11 February 1944, Box 12017, UD 1940–49, RA.

12 The Moscow Embassy to the Foreign Ministry, 29 September 1945, Box 12017, UD 1940–49, RA.

13 Unsigned Foreign Ministry memoranda 15 March 1946, Box 12017, UD 1940–49, RA.

14 Set out in a long memo by the Board of DNL, "Fremstilling av hovedtrekkene i Det Norske Luftfartsselskap A/S's utvikling 1946–1948 og forslag til retningslinjer for selskapets fremtidige virksomhet," 14 January 1949, 124, Box "DNL's utvikling frem til SAS – Diverse brev og dokumenter 1945–47," SAS Museum Archive.

15 The Norwegian embassy to the Foreign Ministry, 15 May 1946, Box 12017, UD 1940–49, RA.

16 The Embassy in Stockholm to the Foreign Ministry, 25 January 1954, Box 3563, Archival Code 55.15, UD 1950–59, RA, with the Swedish press release of a 1954 amendment to the 1946 agreement. Strangely, this agreement is not mentioned in the Foreign Ministry archives for 1940–49. The ongoing Swedish and Danish negotiations (which happened at the same time as the Norwegian ones) were noted, but the last reference to Swedish negotiations, in a message from the Norwegian embassy to the Foreign Ministry on 29 June 1946, was that the Swedish delegation were deeply offended, having left Moscow after five weeks without having met the Soviet negotiators once.

17 *Norge og Sovjetunionen 1917–1955: En utenrikspolitisk dokumentasjon*, ed. Sven G. Holtsmark (Oslo: Cappelen, 1995), "Excerpt from the USSR ambassador to Norway N. D. Kuznetsov's report on Norway's strategic importance and USSR policy towards Norway," June 8, 1946, at 376–79; and ibid., "Excerpt from the USSR ambassador to Norway N. D. Kuznetsov's correspondence to Foreign Minister V. M. Molotov on Norway's relationship with USA and Great Britain and USSR policy towards Norway," May 5, 1947, at 403–4.

18 Knut Einar Eriksen and Helge Øystein Pharo, *Norsk Utenrikspolitikks historie, v: Kald krig og internasjonalisering 1949–1965* (Oslo: Universitetsforlaget, 1997), 197.

19 Stian Bones, "Med viten og vilje: Einar Gerhardsens reise til Sovjetunionen i 1955," *Nytt Norsk Tidsskrift*, no. 3 (2006): 286.

20 Minutes of the meeting of the Extended Foreign and Constitution Committee of the Norwegian Parliament, March 23, 1957, www.stortinget.no/globalassets/pdf/storting sarkivet/duuk/1946–1965/570323u.pdf.

21 Finn Skartum, Head of the Foreign Ministry's 5th Office of Trade Policy, memo, 23 November 1955, Box 3542, UD 1950–59, RA.

22 Head of the Finnish negotiating team to the Head of the Soviet team, copy, 22 October 1955, Box 3542, Archival Code 55.4, UD 1950–59, RA. The letter confirmed that an agreement between Finland and the USSR had been reached on 19 October 1955, and the full agreement was attached. It is not clear how this came into the hands of

the Norwegian Foreign Ministry. In the same archive box is an undated copy of the Austrian–Soviet agreement of 5 November of the same year (1955 being the year when the joint Allied–Soviet occupation of Austria ceased and Austria regained its independence, which may be why Austria was the first country after Finland to obtain an aviation agreement with the USSR).

23 The Embassy to the Foreign Ministry, 19 October 1955, Box 3563, UD 1950–59, RA, to report Soviet press reaction to the agreement with Finland.

24 The Embassy to the Foreign Ministry, 1 October 1953, Box 3563, UD 1950–59, RA.

25 The Embassy in Stockholm to the Foreign Ministry, 25 January 1954, Box 3563, UD 1950–59, RA, enclosing a Swedish press release about the agreement.

26 Ambassador Braadland in Moscow to the Foreign Ministry, 4 February 1956, Box 3542, UD 1950–59, RA, reporting the Swedish–Soviet negotiations.

27 Finn Skartum, memo, 23 November 1955, Box 3542, UD 1950–59, RA.

28 Nils Langhelle had been Norwegian Transport Minister in 1946, when he signed the initial agreement establishing SAS Overseas. In 1952–1954 he was Defence Minister, before he became a Norwegian Labour Party MP and Head of SAS Norwegian Division in August 1955.

29 Finn Skartum, memo "Rettigheter for SAS i Sovjet-Samveldet," 28 November 1955, Box 3542, UD 1950–59, RA.

30 Finn Skartum, memo "Luftfartsforhandlinger med Sovjet-Samveldet," 31 January 1956, Box 3542, UD 1950–59, RA.

31 "Vil Sovjet Springe SAS-samarbeidet?" *Aftenposten,* February 8, 1956.

32 Finn Skartum, memo "Rettigheter for SAS i Sovjet-Samveldet," 9 February 1956, Box 3542, UD 1950–59, RA, enclosing a memo from the Swedish embassy in Oslo.

33 "Svenskene skal fly til Moskva," *VG (Verdens Gang)*, February 9, 1956.

34 "SAS og Moskva," *VG,* February 9, 1956.

35 Finn Skartum, memo "Luftfartsavtale med Sovjet-Samveldet," 10 February 1956, Box 3542, UD 1950–59, RA. It ends with Skartum's handwritten note that the Paris Delegation had been asked to inform NATO.

36 The Swedish Embassy in Oslo to the Norwegian Foreign Ministry, 14 February 1956, Box 3542, UD 1950–59, RA, with a 54-page report on the negotiations from 27 January to 7 February.

37 The Embassy in Moscow to the Foreign Ministry, telegram, 15 February 1956, Box 3542, UD 1950–59, RA.

38 Foreign Ministry press release, 24 February 1956, Box 3542, UD 1950–59, RA.

39 Foreign Minister Hallvard Lange presentation to the Council of Ministers, 9 March 1956, Box 3542, UD 1950–59, RA.

40 The Embassy in Moscow to the Foreign Ministry, 23 March 1956, Box 3542, UD 1950–59, RA. Includes a handwritten note that the Foreign Minister gave ambassador Braadland leave to sign the agreement

41 Contrary to aviation agreements of the kind preferred in Europe today, which usually do not go into operational details, this agreement was intended to regulate operations very closely, the specifics were integrated either into the agreement or (if to be agreed) into the sub-agreements between the two preferred airlines, DNL (later SAS) and Aeroflot.

42 Agreement between SAS and Aeroflot, 31 March 1956, Box 3542, UD 1950–59, RA.

43 The Ministry of Transport to the Foreign Ministry, 2 November 1956, Box 3542, Archival Code 55.4, UD 1950–59, RA, asking the Foreign Ministry to secure this alternate route in the agreement. For Soviet intransigence, see Ambassador Braadland in Moscow to the Foreign Ministry, 26 June 1957, Box 3543, UD 1950–59, RA.

44 The Norwegian delegation to the Scandinavian aviation negotiations with the Soviet Union, report for the period 21–31 March 1956, 5 May 1956, Box 3542, UD 1950–59, RA: 9.

45 Foreign Ministry memo 31 March 1958, Box 3542, UD 1950–59, RA.
46 Minutes of the meeting of the Extended Foreign and Constitution Committee of the Norwegian Parliament, March 23, 1957, www.stortinget.no/globalassets/pdf/storting sarkivet/duuk/1946–1965/570323u.pdf, with foreign minister Hallvard Lange's account of what the Soviet ambassador said at dinner.
47 The Swedish Embassy to the Norwegian Foreign Ministry, 8 February 1956, Box 3542, UD 1950–59, RA, asking for Norway's view on entering into an aviation agreement with the Soviet Union modelled on the Austrian–Soviet agreement of the autumn of 1955.
48 Hjalmar Riiser Larsen, *Over Nordpolen*, 8 mm film (Oslo: SAS/Norsk Dokumentarfilm, 1958). The 18-minute-long film has been digitized and can be viewed at the Norwegian National Aviation Museum.
49 Agreement between Norway and the Soviet Union concerning aviation, March 31, 1956, https://lovdata.no/dokument/TRAKTAT/traktat/1956-03-31-1.
50 In the updated agreement on lovdata.no, the name change is annotated as "The agreement is carried on with Russia."
51 *Norges Traktater 1661–1966*, iii: *1956–1967* (Oslo: Det Kgl. Norske Utenriksdepartement, 1968), 47 and no. 1. See also the Transport Minister's presentation to the Council of Ministers, 3 February 1967, Box S-1713/D/Db/L0122, Folder 03: Luftfartsforhandlinger Sovjet 1967–68, Ministry of Transportation Archive (SD), RA.
52 Ibid., 4.
53 The Foreign Ministry to the Transport Ministry, the Defence Ministry, the Norwegian Civil Aviation Authority, and the Norwegian Embassy in Washington DC, 8 November 1966, Box S-1713/D/Da/L0069, SD, RA, reporting on a visit to the Foreign Ministry by the US Chargé d'affaires in Oslo, Mr Bovey. Mr Bovey delivered a US memo asking Norway to show understanding for the US view on Cuba and reject Soviet demands for transit rights to Cuba via Norwegian airspace.
54 The Transport Minister's presentation to the Council of Ministers, 11 December 1970, Box S-1713/D/Db/L0124, SD, RA. The negotiations were prolonged – 5 rounds of negotiations between 24 November 1969 and 2 October 1970 – and the agreement was signed by the Soviet Aviation Minister B. P. Bugajev and the Norwegian Transport Minister Håkon Kyllingmark when Bugajev visited Oslo on 27–30 January 1971. The official records in Lovdata has the agreement, which was registered on 11 February, as effective from 19 May 1971, but SAS was allowed to start operations the month before.
55 Anders Buraas, *Fly Over Fly: Historien om SAS* (Oslo: Gyldendal, 1972), 271.
56 "Agreement Between the Government of Norway and the Government of the Union of Socialist Republics Regarding Development of the Trans-Siberian Route, 10 April 1981," in *Overenskomster med fremmede stater 1981* (Oslo: Det Kgl. Utenriksdepartement, 1982), 693.
57 Interview with Øyvind Ek (Head of the Aviation Section in the Norwegian Transport Ministry) by Karl L. Kleve, Oslo, Norway, 18 October 2017.

References

Bones, Stian. "Med viten og vilje: Einar Gerhardsens reise til Sovjetunionen i 1955." *Nytt Norsk Tidsskrift*, no. 3 (2006): 276–86.
Buraas, Anders. *Fly Over Fly: Historien om SAS*. Oslo: Gyldendal, 1972.
Eriksen, Knut Einar, and Helge Øystein Pharo. *Norsk Utenrikspolitikks historie, v: Kald krig og internasjonalisering 1949–1965*. Oslo: Universitetsforlaget, 1997.
Haanappel, Peter P. C. *Ratemaking in International Air Transport*. Deventer: Kluwer, 1978.
Holtsmark, Sven G. *Norge og Sovjetunionen 1917–1955: En Utenrikspolitisk Dokumentasjon*. Oslo: Cappelen, 1995.

Papatheodorou, Andreas. "The Impact of Civil Aviation Regimes on Leisure Travel." In *Aviation and Tourism*, edited by Anne Graham, Andreas Papatheodorou, and Peter Forsyth, 49–57. Farnham: Ashgate, 2008.

Rosenberg, Arne. *Air Travel within Europe*. Stockholm: National Swedish Consumer Council, 1970.

Strauss, Anselm. *Negotiations: Varieties, Contexts, Processes and Social Order*. San Francisco: Jossey-Bass, 1978.

Tjomsland, Audun, and Kjell G. Wilsberg. *Mot Alle Odds*. Oslo: Braathens SAFE, 1995.

10 Concluding remarks

Tourism across a porous curtain

Angela Romano

This book focuses on tourism behind the Iron Curtain. Cold War Europe comprised not only Europeans and their countries, but also the superstructure in which they lived, namely the bipolar divide, two military alliances facing each other, and the leadership and deep involvement of two extra-European superpowers that had interests and quarrels at the global level. However, as this chapter will highlight, Cold War Europeans were capable of developing transcontinental dynamics that differed from and transcended the superpower bipolar relationship and its ups and downs, challenged the bipolar divide, and gradually yet steadily promoted a new kind of thinking on the Continent based on webs of bilateral and multilateral cooperation. Tourism, it will be shown, became part and parcel of this process of pan-European cooperation, as well as the expression – both East and West – of ideological and political visions of international relations, economic interests, strategies of growth, and regimes' self-confidence (or the lack thereof). Consequently, this chapter will also argue in favour of new avenues of research which, by taking tourism as a heuristic tool, will contribute to a more sophisticated understanding of the Cold War's evolution and end in Europe.

Europe as the template for the Cold War – and for its overcoming

A few years ago, discussing the ever-expanding scope of Cold War historiography, the historian Federico Romero made a strong case for "re-emphasis(ing) the place of Europe in the global Cold War."[1] He noticed that the Cold War's paradigms and defining features were conceived for application to the European theatre first: territorial partition; socioeconomic separation; alliance systems with vast military structures; intra-bloc institutionalised economic interdependencies; and vigorous ideological confrontation, shaping cultural representations and mobilising civil society.[2] More importantly, he remarked, the Cold War originated in and about Europe, "pivoted on the continent's destiny," and found its solution in Europe.[3]

We may add that Europe was central to Cold War symbolism. The Cold War imposed a mental mapping that was characterised by the idea of otherness as a necessarily antagonistic entity. Europe first and foremost was framed as a space shaped by a dualistic concept of *us* and *them*: East or West, backward or progressive, dictatorial or democratic, repressive or free – or vice versa, solidary or exploitative, moral or corrupted, fostering brotherhood or promoting individualism. That the structure, features, and constructed views of the Cold War found perfect expression in the partition of Europe also explains why Churchill's early Iron Curtain image – a European-based image – endured down the decades as the most powerful symbol of the Cold War worldwide.

In addition to confrontation being a defining element of the Cold War, we must acknowledge isolation as one of its key features. This is visible not only in the military, political, and economic organisation of the blocs, but also in the regimes' attempts to obstruct possible contamination by the ideas of the other camp. Here again, the importance of the European reality – with the Berlin Wall as perfect epitome – is crucial also in cultural and symbolic terms. This view is confirmed by the historiography of post-Cold War Central and Eastern Europe, which offers a narrative of these countries' "return to Europe." The fact that the forty-year-long socialist experience is presented as an interlude in an otherwise all-European or pan-European history only strengthens the image of a Cold War Europe in which the Iron Curtain was very much present as a physical, ideological, and even psychological barrier, secluding people from economic, social, and cultural contamination as well as mere contacts with the other side.

The Cold War-era partition left a legacy of separate studies of Eastern and Western Europe; historians working on the two sides of what used to be a divided Europe have proceeded with largely separate agendas and networks. In the last decade, however, an ever-growing number of scholars has focused on East–West relations in Europe during the Cold War, putting Europeans' agency centre stage. This flourishing historiography is deeply changing our understanding of the continent as the realm of confrontation and separation of the two ideological systems. By re-focusing their attention on Europe in the global Cold War, historians are adding layers of complexity to our understanding of East–West relations and leading to a more sophisticated assessment of the Cold War as a historical process.

Recent studies recognise the 1970s in particular as the period in which the geopolitical and ideological bipolar equilibrium eroded, and small and medium powers enjoyed greater autonomy from the superpowers.[4] In this context, studies of detente have proved that the latter had a substantially different meaning for the superpowers and their allies. While the former intended detente as a means to consolidate bipolarity and lower the costs and risks of superpower confrontation, "European detente" was meant to promote a gradual overcoming of the Cold War in Europe. This process of rapprochement between Western Europe and the socialist countries was to be achieved through expanding contacts and deepening mutual interdependence between the two halves of the continent.[5] Indeed, most Western European governments' policies of detente deliberately involved the East in commercial, financial, and cultural cooperation. This European detente is now acknowledged among the crucial factors in determining the end of the Cold War, and explaining the pace of the fall of communism in Europe.[6]

What is becoming clearer is that a complex and lasting pattern of European detente can be counted among the key features of the Continent from the mid-1960s until the end of the Cold War. Since the mid-1960s, most Western European governments promoted, through bilateral channels, a more or less successful policy of detente with the Soviet Union and the Eastern European countries.[7] By the mid-1960s the socialist regimes recognised foreign trade as an important factor in socialist economic development, and planned to expand trade with the developed market economies.[8] Consequently, the socialist regimes of Europe grew ever more enmeshed in trade, finance, and exchanges with capitalist Western Europe.

While still in place in the ideological, security and symbolic spheres, the Cold War partition of the Continent was becoming less stringent. East and West remained separate and antagonistic camps, but they were connected by multilateral and bilateral patterns of interaction. By the mid-1970s, Europe was crisscrossed by an expanding web of exchanges that prefigured an area of pan-European cooperation, which was often read with an expectation of gradual convergence and interdependence. This emerging continental space of collaboration was also enshrined in the 1975 Final Act of the Conference on Security and Cooperation in Europe (CSCE). Moreover, the economic, diplomatic, societal, and cultural connections that had come to define detente between Eastern and Western Europe did not wither in the late 1970s and early 1980s, a period that several historians (who mostly focus on the superpowers) still label the "second Cold War." Very recently, Oliver Bange and Poul Villaume have argued strongly against such notion and pointed to the continuity and relevance of a long detente, which they define as "antagonistic cooperation" with strong elements of a "trans-bloc, trans-societal, and trans-ideological framework" with European actors at its centre.[9]

Recent European Cold War historiography is paying much-deserved attention to the role of neutral countries in what is therefore confirmed as a complex and multifaceted space featuring not only East–West rivalry, but also diverse interactions and pan-European cooperation. In addition, some historians have recently proved that this opening pan-European space also invited the action of actors that were previously insulated or passive, such as the European Community. Since the early 1970s, the enlarged, strengthened, and more politically active European Community had a vested interest in the continuation of detente and the promotion of new European relations beyond the Cold War blocs' antagonism.[10] More importantly, the European Community proved not only willing but also able to significantly alter intra-European relations, cutting many of the blocs' ties in the East.[11]

Recent historiography has therefore demonstrated that, in addition to confrontation, Cold War Europe experienced a growing degree of East–West connectedness and interdependence. The change of focus is also visible in the titles of the literature. Some historians emphasise the intentionality of the promotion of these contacts, conceived as a means to change the European order in the long run, hence the likes of *Overcoming the Cold War*; *Helsinki 1975 and the Transformation of Europe*; *Perforating the Iron Curtain*; *Overcoming the Iron Curtain*; and *Untying Cold War Knots*.[12] Others have been more interested in giving prominence to the development of multiple and diverse contacts across the Cold War divide, as in *Raising the Iron Curtain*; *The Nylon Curtain*; *Passing Through the*

Iron Curtain; *Gaps in the Iron Curtain*; *The Iron Curtain as a Semi-permeable Membrane*; and *Loopholes in the Iron Curtain*.[13]

Overall, this impressive historiographical production offers clear evidence that a diverse and numerous group of predominantly European actors were proactive in encouraging, building, and effecting contacts and exchanges across Cold War borders, and later were committed to preserving this web of relations from the harsh winds of renewed superpower confrontation. They confirm the argument that Cold War Europe's two antagonistic camps were divided by a porous curtain rather than an iron one.

It is also becoming evident that European detente paved the way for new think-ing and deeper cooperation across the Continent. This development is epitomised by the Helsinki CSCE and its ensuing process, which Cold War historiography now recognises as having had a key role in bringing about the fall of socialism.[14] A burgeoning scholarship on the CSCE in the past decade has demonstrated that it was of major importance in most states' Cold War policy, which has contributed to elucidate the different conceptions of détente and to reveal the relevant role and increasing activism in Europe of actors other than the superpowers. The analysis of the CSCE negotiations and its Final Act reveals that the pan-European confer-ence was a step towards overcoming the Cold War order's logics and constraints.[15] In particular, it has demonstrated that the Helsinki process was a key instrument in Western European and neutral states' detente policies, as well as in the European socialist governments search for a more autonomous role in a new framework of multilateral cooperation.[16]

More recently, a group of historians teamed up to offer the first multifaceted analysis of the increasingly relevant yet contradictory place that pan-European space occupied in the economic and political life of socialist regimes. They have identified common patterns across the socialist bloc, but also the rifts over the desirability or necessity of opening up to international exchange.[17] An even more ambitious research project, led by the same historians, is now exploring the changing mindset of the European socialist elites when cooperating with Western Europe and the EEC, and the existence of a plurality of views in each country. The project reconstructs and assesses the expectations that nurtured the social-ist ruling elites' approaches to the international division of labour and European cooperation, their national strategies across the 1970s, their attempts to reconcile transformation with regime stability, and ideological rivalry with a new rhetoric of collaboration – and the predicaments the socialist regimes faced as their strate-gies began to unravel.[18]

The emphasis on improved East–West contacts and cooperation is not to deny the persisting reality of Cold War antagonism. Control and limitations were still in place or put in place by socialist regimes to respond to the proliferation of con-tacts through an all-too-porous curtain. It is enough to remember the poor record of most socialist regimes in implementing the CSCE provisions pertaining to the improvement of citizens' rights to access Western territory, literature, and the press, as well as their jamming of foreign radio broadcasts. Another key example has been the impressive growth in staff and activities of the security apparatuses since the 1970s, which developed new justifications for mass surveillance precisely because of the policy of detente and increasing contacts between East and West, which the agencies saw as a fundamental threat from hostile influences.[19]

Yet people did travel and encounter "the others." The very fact that people were being allowed to travel behind the Iron Curtain could be considered a twofold achievement, as an improvement in domestic regimes' relations with their citizens *and* as a barometer of improved East–West relations. This was epitomised, once again, in the Final Act of the Helsinki CSCE, where tourism was linked to actions intended to favour freer movement across borders, and taken as one of the yardsticks of the governments' commitment to detente and pan-European cooperation.

Tourism in the framework of the CSCE

The CSCE Final Act was a politically solemn but non-legally binding agreement, which comprised three main sets of recommendations (the so-called baskets): (I) questions relating to security in Europe (comprising ten principles guiding relations between the participating states, known as the Helsinki Decalogue, as well as the Confidence Building Measures); (II) cooperation in the fields of economics, science and technology, and the environment; and (III) cooperation in humanitarian and other fields.

The inclusion of Basket III was entirely a Western idea and diplomatic victory. It endorses the liberal concept of human rights and centrality of the individual, and hence reversed the Soviet view, according to which detente only related to relations among states. The Final Act gave governments and dissidents an opportunity to legitimately claim the modification of certain rules and practices of socialist regimes towards their own citizens. This was a main change in international law, as it asserted the idea that the way states treat their citizens was now a matter of international jurisdiction. Indeed, the West's emphasis on human contacts was justified in terms of a "people first" approach to detente, which also applied to proposals in the field of economic cooperation.[20] It is in relation to this "human contacts" aspect that tourism features in both Baskets II and III.

The West wanted tourism in Basket II because of the tie-in with freedom of movement.[21] The European Community member states presented a common draft recommendation to the CSCE on 28 January 1974, which became the basis for the negotiations. Essential EC proposals included the facilitation of tourists' mobility within the country visited as well as larger currency allowances for travel abroad. Both were first proposed by the Italian government, which specifically highlighted their relevance beyond economic and commercial concerns into the social field and human relationships. There were also more economic rationales: another Italian proposal called for an in-depth study of the statute and the activities of travel agencies, while a joint Irish–Italian proposal asked to pay more attention to staggering holidays in order to avoid excessive concentration of tourists in the summer season. Following careful consideration at European Community level, these proposals were brought together as a draft recommendation submitted to the CSCE. Much of the editorial work was completed in the spring of 1974. Only a few paragraphs remained problematic, namely those on facilitating individual and group tourist movement with the possibility of obtaining documents and foreign currency to travel (the socialist regimes' delegations were firmly opposed to the discussion of these issues, which they considered a matter for Basket III); and on the activities of foreign travel agencies, which the socialist countries did not want to be specifically listed. Agreement was finally reached in the second week of December 1974, with the EC member states settling for a less constraining wording. Instead of "currency," which was an unacceptable term to socialist regimes, the text spoke of "financial means," and now included the caveat that individual countries' economic possibilities should be taken into account. The reference to the granting of documents was eventually phrased as "the necessary formalities for travel."[22]

The other Basket II provisions concerning tourism were more focused on economic aspects. Cooperation in economic fields represented the second major topic of interest for the socialist countries (the first being security), and this helped the work on Basket II proceed quite fast. In addition, an ad hoc coalition of Southern European countries – Portugal, Spain, Italy, Yugoslavia, Greece, Turkey, Romania, and Bulgaria – was particularly active in the sub-commission dealing with the promotion of tourism, on which their economies were clearly dependent.[23] States intended to increase tourism by "encouraging the exchange of information, including relevant laws and regulations, studies, data and documentation relating to tourism, and by improving statistics with a view to facilitating their comparability," and by "facilitating the activities of foreign travel agencies and passenger transport companies in the promotion of international tourism." However, the West could not get a provision that engaged socialist countries to allow private agencies to advertise and operate normally in socialist countries' territory.[24] States also agreed to engage to "pursue their cooperation in the field of tourism bilaterally and multilaterally with a view to attaining" specific objectives such as improving tourist infrastructure, examining possibilities of exchanging tourism specialists and students with a view to improving their qualifications; promoting conferences and symposia on the planning and development of tourism, and encouraging tourism outside the high season. They also pledged to "endeavour, where possible, to ensure that the development of tourism does not injure the environment and the artistic, historic and cultural heritage in their respective countries."[25]

The most difficult negotiations on tourism took place in the highly contentious Basket III, where the Soviet and their allies were determined to subject all provisions to the principles of sovereignty and non-interference in internal affairs.[26] For the West, tourism was a peculiar aspect of the human contacts and freer movement issue.[27] During the 1960s the percentage of tourism from the West to the East had significantly increased, while the opposite flow had remained at its negligible level. In most socialist countries – the worst case being the Soviet Union – tourists to the Western world went through complex and arbitrary procedures. The authorities essentially promoted collective tourism, which enabled them to exercise effective control over the actions of tourists; pre-established programmes and itineraries allowed the regime to limit private contacts with the foreign population as well as the risks of defection. Foreigners visiting socialist countries faced considerable restrictions on movement as well as on contact, direct or indirect, with local citizens. Socialist regimes also had an interest in limiting the number of citizens exposed to the wealth of Western societies. Moreover, given the lack of hard currency in the socialist bloc, governments considered themselves justified to impose restrictions to their citizens willing to travel abroad. In addition, tourists were only allowed to carry small sums of money. Conversely, socialist regimes maintained abnormally high exchange rates for foreigners coming to visit the country, in order to exact more hard currency.[28]

The overriding Western preoccupation was to draw the socialist delegates into a serious discussion of measures that would have practical and discernible effects on the circulation of people (and information) between East and West. The EC member states made it clear that only if satisfied with the Basket III provisions would they accept to move to the CSCE final stage and to hold it at the summit level (which was a priority for the Soviets). On the question of the freedom of movement, the West asked for the removal of impediments upon travel of Eastern European citizens to non-socialist countries, for example the reduction of passport fees, abolition of exit visas requirements in conformity with general practice in the West, liberalisation of foreign exchange allowances, simpler and more transparent administrative procedures for visa requests, and the possibility to appeal in case of denial or undue delay. The citizens of the participating states should be permitted and encouraged to travel to and within the other countries in Europe, and they should suffer no adverse effects for applying.

The Eastern delegations argued that these problems in many cases could not be usefully discussed at the Conference and should be solved bilaterally. They also stressed the differences between the Eastern and Western political and social systems, and the need for scrupulous observation of the principles of non-intervention and respect for domestic laws and customs.[29] The socialist countries took the line that the whole of Basket III should be governed by a preamble whose wording was designed to provide them room for maintaining existing restrictive practices (and freedom to introduce new ones); the view was clearly that the more detailed the substantive provisions on human contacts and information, the more explicit the restrictive references in the preamble should be.[30]

Work on the preamble and provisions related to travel formalities remained deadlocked for months.[31] By the end of April 1975 not a single word had been registered, and attitudes had hardened on all sides on the politically sensitive issues of working conditions for journalists, access to information, freer travel, and the general objectives for human contacts and information.[32] In mid-May 1975 a package deal prepared by the British, approved by all EC partners, and supported by other NATO allies and neutral countries met a more forthcoming attitude from the socialist delegates. The contents of the package did not of course match up to all the ambitions of the Western governments, yet it represented a satisfactory outcome.[33] With regard to human contacts, the states made it "their aim to facilitate freer movement and contacts, individually and collectively, whether privately or officially, among persons, institutions and organisations of the participating states, and to contribute to the solution of the humanitarian problems that arise in that connection."[34] The formulation was neatly Western, as it explicitly mentioned individual and private contacts and movements.

Tourism features in two specific items under the "human contacts" rubric: "Travel for Personal or Professional Reasons," and "Improvement of Conditions for Tourism." Overall, the provisions tackled administrative hindrances and were meant to reduce the chances of a person being penalised for trying to travel abroad.[35] The Western European countries only obtained two provisions of general intent. First, participating states agreed to endeavour to lower, where necessary, the fees for visa and official travel documents; as it is plain to see, the states had wide discretion in determining individual cases. Second, states declared their intention to ease regulations concerning movements of foreign citizens within their territory, with due regard to security requirements; yet the West could not gain free movement of foreigners in one state's territory apart from in identified security areas. Probably, the major gain in this field was the specific clause on contacts and meetings among religious faiths, institutions, and organisations, which was vigorously put forward by the Vatican delegation and supported by the Italians.

Despite undeniable limits and weaknesses, Basket III offered an overall framework for intergovernmental cooperation and a series of guidelines to participating states for unilateral implementation of reforms or arrangements to comply with their international political undertakings. Moreover, the Final Act set in motion a process, or at least the first step of a process, by calling for the convening of a follow-up meeting in Belgrade in two years' time, in order to check the implementation of Final Act provisions and to promote further cooperation. Other meetings followed in subsequent years, turning the CSCE into the Helsinki process.[36] Of course the Final Act did not bring about a massive liberalisation of travel. Yet in the ensuing years various bilateral agreements were concluded, most often on a reciprocal basis, to facilitate travel and to establish cooperation in the field of tourism. Moreover, various socialist countries took a series of unilateral measures to ease the conditions of entry and temporary exit for family visits and tourism. For instance, in 1977 both Hungary and Bulgaria abolished the obligation on Western tourists to change a certain daily amount of foreign exchange, and in 1978 Bulgaria adopted the application of preferential exchange rates or tourist tariffs. In 1977 Bulgarian authorities also approved the granting of entry and transit visas on arrival at the borders. In Poland, a decision in 1982 reduced visa deadlines, extended the validity of passports to three years, and required written reasons to be given when visa applications were refused.[37] To some extent, the Western Europeans' CSCE promotion of human contacts across the blocs did affect the way socialist regimes treated foreigners and their own citizens.

Avenues of research

The CSCE case illustrates the strong connection that governments in both East and West (as well as the neutral ones) established between tourism and travelling on the one hand and Cold War politics on the other. It also hints at specific economic interests that could cut across the East–West divide and foster ad hoc transversal alliances. Moreover, the CSCE, though establishing a multilateral framework, asserted the crucial importance of action taken at the bilateral level as well as in domestic policy. Lastly, the Final Act emphasised the importance of allowing the people to meet and know "the others" across the whole continent. It is possible to identify three levels of entanglement between tourism and the Cold War – international, domestic, and personal – or, in other words, relations among states; relations between the state and its own citizens as well as foreign tourists; and the experience of the tourists behind the Iron Curtain.

Not only does tourism offer a unique perspective that further elucidates the multifaceted phenomenon of East–West relations, it also opens another window into socialist regimes' foreign, economic, and domestic policies. The chapters in this volume address a variety of important issues that open up new avenues for research, linking the histories of tourism and of the Cold War in meaningful ways. They also further encourage collaborative, cross-feeding efforts at research and conceptualisation by historians working on various spheres of European and international history, communism, economic, social, and cultural history.

Individual or group contacts through travelling and tourism amount to a transnational activity that is often difficult to trace and even more difficult to interpret. A micro-history perspective per se adds to our knowledge, but has a broader significance if connected with larger events and processes, such as Cold War relations, regional cooperation, and domestic revolutions. This approach necessarily requires a certain degree of quantification and qualification of the tourism and travel experience. How large was the observed phenomenon in specific country or countries under scrutiny? Were tourists and travellers a relevant part of the population at the given time and/or in comparison with other periods? Even more important is to detect and appraise the profile and background of tourists and travellers. Was the activity of travelling spread across the strata of the population – countrymen versus city-dwellers, apparatchiks versus workers? Were Western tourists and travellers mostly leftists – if not fellow travellers, at least left wing? How many Western tourists were connected to the East via family or other bonds from the pre-communist period?

Profiling tourists and travellers would certainly help us to better assess their openness to new perceptions of the country and society they visited. Tourism and travelling are often identified with the desire to discover the "other" and have often been adopted as a tool for improving relations between countries. Yet travellers have their own pre-constructed views of the other, and thus might not be open to "discovering" and changing their existing interpretative paradigms. A real and direct experience can even strengthen pre-set stereotypical images of the country and people visited. In an era and space dominated by Cold War antagonism and pervasive indoctrination of the masses on the superiority of one's own system and the backwardness or even evil nature of the other system, how much was the travelling and tourist experience influential in changing their views of the host country? And did perceptions of the other influence the tourists' allegiance to their own system? For example, Western workers who in the 1970s felt the effects of the economic crisis might have been more sensitive to low-price services offered in socialist countries.

All of the preceding suggests a significant degree of instrumentality of tourism and travelling across the Iron Curtain. Future research could well focus on identifying actors for whom tourism and travelling was an instrument to reach goals other than personal leisure of the travellers (the variety of actors can be impressive: not only governments, but also NGOs, activists, associations); on appraising their goals, and assessing the results. This volume shows that we may record cases going in very opposite directions, namely tourism as a means to transcend Cold War conflicting views or travels meant to strengthen the regime's self-constructed image of superiority and showcase the regime's achievements. Likewise, when assessing the results of using tourism as a means to achieve a specific goal, research may uncover cases in which carefully devised tourist and travelling experiences proved useful, others in which they were irrelevant, and other cases in which they turned out to be counterproductive and left the tourists with a poor impression.

Another crucial field of historical enquiry relates to the agencies responsible for tourism policy and its implementation, and the exploration of their interplay. Although there was space for contacts at unofficial level, the state apparatus remained the main actor responsible for encouraging or limiting these connections via regulations of various kinds, allocation of financial support, planning and building of infrastructure, recruitment in the tourist sector, and of course direct control. First, research linking tourism and travel to socialist regimes studies contributes insights on socialist decision-making and the specific role of specialised organs of the government. While there is a rich literature on the Soviet Union and what Alec Nove defined as "centralized pluralism," research on bureaucracy and interest groups in other socialist countries is still scant and rarely goes beyond the 1960s. A diachronic inquiry into the actors involved in the state apparatus dealing with tourism and travelling and into the rules regulating the sector would evidence the impact of generational change or ideological turns, shed additional light on the regime's approach to both domestic reforms and foreign relations, and relate it to historiographical debates on the periodisation of the Cold War and detente.

Second, the study of state regulations and practices pertaining to travel and tourism can foster an understanding of the complexities of the socialist world, as it brings additional evidence of a diverse range of approaches to relations with the capitalist West (as well as to relations with fellow socialist countries). To take just one example, the Polish government after Gierek came to power adopted a relatively liberal passport policy, thanks to which the number of visits abroad skyrocketed from approximately 1 million in 1971 to 10 million in 1972. The policy remained in place without major interruption until the imposition of martial law in December 1981.[38] Poland's liberal passport policy stands in sharp contrast to the Romanian or Soviet rules and vetting practices for citizens travelling abroad. At the same time, it is worth analysing and assessing the different regimes' regulations and practices for incoming Western tourists, as it was often feared they would spread the "germs" of capitalist views, morals, and mentality, at the very least to and through the tourism workers with whom they had actual and prolonged contact.

The degree of distrust, control, and limitations imposed on Western tourists speaks volumes about the self-confidence of the various socialist ruling elites. Research on tourism and travel that takes into consideration the relationship between travellers and those in charge of setting the rules helps assess how the socialist regimes' saw their country's place and prospects in an emerging space of trans-European connections that challenged their political control and ideological legitimisation.

Third, this research also sheds new light on the fabric of socialist societies, on the progressive loss of citizens' allegiance, and on the inherent weaknesses and eventual fall of the socialist system. As Péteri suggests, the socialist regimes' relationship with the West lay at the core of their identity and self-understanding, given the fundamental claim that socialism was constructing a superior, alternative modernity.[39] There was no other part of the world with which sentiments of inferiority and superiority, admiration and enmity, emulation and rejection became so intertwined. Ultimately, the West was not only a rival, but also an inextricable part of the fabric of socialist society.[40] The West, in the evocative title of Paulina Bren's 2008 article, was the "Mirror, mirror, on the wall," a means of judgement on oneself.[41] In this respect, tourism and travelling certainly features among the various types of East–West interactions and the many layers of the socialist regimes' engagement with the West.

Historians working on tourism and travel who take a transnational perspective thus enrich both the national historiographies of socialist regimes and international history studies. Transnational historians can reconstruct and reveal the diverse geographies of economic, social, political, and intellectual interactions that made Europe a continent of overlapping spaces of cooperation rather than a place hosting clearly circumscribed and isolated systems of state socialism on the one hand and capitalism on the other. Péteri talks about the "nylon curtain," which

> was not only transparent but it also yielded to strong osmotic tendencies that were globalizing knowledge across the systemic divide about culture, goods, and services. These tendencies were not only fuelling consumer desires and expectations of living standards, but they also promoted in both directions the spreading of visions of . . . civil, political, and social citizenship.[42]

The historian Arnd Bauerkämper affirms that "Altogether, the history of Europe is to be conceived as the history of continuous social and cultural exchange, interaction, and networking."[43] In this respect, the history of tourism and travel enhances our understanding of the history of Cold War Europe as a place where connectedness came to characterise the continental order, and tourists crossed what was in fact a porous curtain.

Notes

1   Federico Romero, "Cold War Historiography at the Crossroads," *Cold War History* 14, no. 4 (2014): 687.
2   Ibid.; see also Melvyn P. Leffler, *For the Soul of Mankind: The United States, the Soviet Union, and the Cold War* (New York: Hill & Wang, 2007).

3 In Federico Romero, *Storia della guerra fredda: L'ultimo conflitto per l'Europa* (Turin: Einaudi, 2009), Federico Romero convincingly argues that the Cold War was the clash between two new world powers just as much as the last conflict for Europe.

4 Odd Arne Westad, ed., *Reviewing the Cold War: Approaches, Interpretations and Theories* (London: Frank Cass, 2000); Franz Knipping and Matthias Schönwald, eds., *Aufbruch zum Europa der Zweiten Generation: Die europäische Einigung 1969–1984* (Trier: WVT, 2004); Antonio Varsori, ed., *Alle origini del presente. L'Europa occidentale nella crisi degli anni Settanta* (Milan: Franco Angeli, 2007).

5 Wilfried Loth and George-Henri Soutou, eds., *The Making of Détente: Eastern and Western Europe in the Cold War, 1965–75* (London: Routledge, 2008); Oliver Bange and Gottfried Niedhart, eds., *Helsinki 75 and the Transformation of Europe* (New York: Berghahn, 2008); Angela Romano, *From Détente in Europe to European Détente: How the West Shaped the Helsinki CSCE* (Brussels: Peter Lang, 2009); Andreas Wenger, Vojtech Mastny, and Christian Nuenlist, eds., *Origins of the European Security System: The Helsinki Process Revisited, 1965–75* (London: Routledge, 2009); Poul Villaume and Odd Arne Westad, eds., *Perforating the Iron Curtain: European Détente, Transatlantic Relations, and the Cold War, 1965–1985* (Copenhagen: Museum Tusculanum, 2010); Silvio Pons and Federico Romero, eds., *Reinterpreting the End of the Cold War: Issues, Interpretations, Periodizations* (London: Frank Cass, 2011).

6 Robert English, *Russia and the Idea of the West: Gorbachev, Intellectuals, and the End of the Cold War* (New York: Columbia University Press, 2000); Wilfried Loth, *Overcoming the Cold War: A History of Détente, 1950–1991* (London: Palgrave, 2002); Frédéric Bozo, Marie-Pierre Rey, N. Piers Ludlow, and Leopoldo Nuti, eds., *Europe and the End of the Cold War: A Reappraisal* (London: Routledge, 2008); Romero, *Storia*; Villaume and Westad, *Perforating*; Jussi M. Hanhimäki, "Détente in Europe, 1969–1975," in *Cambridge History of the Cold War*, ed. Melvyn Leffler and Odd Arne Westad (Cambridge: Cambridge University Press, 2010), ii. 198–218; N. Piers Ludlow, "European Integration and the Cold War," in *Cambridge History of the Cold War*, ed. Melvyn Leffler and Odd Arne Westad (Cambridge: Cambridge University Press, 2010), ii. 179–97; John Young, "Western Europe and the End of the Cold War, 1979–1989," in *Cambridge History of the Cold War* (Cambridge: Cambridge University Press, 2010), iii. 288–310; Arne Westad, *The Cold War: A World History* (New York: Basic, 2017).

7 There is rich and ever-expanding literature. See, for example, Maurice Vaïsse, *La Grandeur: Politique étrangère du Général de Gaulle, 1958–1969* (Paris: Fayard, 1998); Marie-Pierre Rey, *La tentation du rapprochement: France et URSS à l'heure de la détente (1964–1974)* (Paris: Publications de la Sorbonne, 1991); Antonio Varsori, *L'Italia nelle relazioni internazionali dal 1943 al 1992* (Rome: Laterza, 1998), 171–98; Loth, *Overcoming*, 89–95; Bruna Bagnato, *Prove di Ostpolitik: politica ed economia nella strategia italiana verso l'Unione Sovietica: 1958–1963* (Florence: L. S. Olschki, 2003); Arne Hofmann, *The Emergence of Détente in Europe: Brandt, Kennedy and the Formation of Ostpolitik* (London: Routledge, 2007); N. Piers Ludlow, ed., *European Integration and the Cold War: Ostpolitik – Westpolitik, 1965–1973* (London: Routledge, 2007); Loth and Soutou, eds., *The Making of Détente*.

8 William V. Wallace and Roger A. Clarke, *Comecon, Trade and the West* (London: Frances Pinter, 1986), 102–3; Ivan T. Berend, "What Is Central and Eastern Europe?" *European Journal of Social Theory* 8, no. 4 (2005): 413.

9 Oliver Bange and Poul Villaume, eds., *The Long Détente: Changing Concepts of Security and Cooperation in Europe, 1950s – 1980s* (Budapest: Central University Press, 2017), 1–15.

10 Angela Romano, "More Cohesive, Still Divergent: Western Europe, the US and the Madrid CSCE Follow-Up Meeting," in *European Integration and the Atlantic Community in the 1980s*, ed. Kiran K. Patel and Ken Weisbrode (Cambridge: Cambridge University Press, 2013), 39–58; Angela Romano, "G-7s, European Councils and East–West Economic Relations, 1975–1982," in *International Summitry and Global*

Governance: The Rise of the G-7 and the European Council, 1974–1991, ed. Emmanuel Mourlon-Druol and Federico Romero (London: Routledge, 2014), 198–222.

11 Angela Romano, "Untying Cold War Knots: The EEC and Eastern Europe in the Long 1970s," *Cold War History* 14, no. 2 (2014): 153–73.

12 Loth, *Overcoming*; Bange and Niedhart, *Helsinki 75*; Villaume and Westad, *Perforating*; Frédéric Bozo, Marie-Pierre Rey, Bernd Rother, and N. Piers Ludlow, eds., *Overcoming the Iron Curtain: Visions of the End of the Cold War in Europe, 1945–1990* (New York: Berghahn, 2012); Romano, "Untying."

13 Yale Richmond, *Cultural Exchanges and the Cold War: Raising the Iron Curtain* (University Park: Penn State University Press, 2003); Gyorgy Péteri, "Nylon Curtain: Transnational and Transsystemic Tendencies in the Cultural Life of State–Socialist Russia and East–Central Europe," *Slavonica* 10, no. 2 (2004): 113–23; "Passing Through the Iron Curtain," special issue, *Kritika* 9, no. 4 (2008); Gertrude Enderle-Burcel, ed., *Gaps in the Iron Curtain: Economic Relations between Neutral and Socialist Countries in Cold War Europe* (Cracow: Jagiellonian University Press, 2009); Michael David-Fox, "The Iron Curtain as a Semi-permeable Membrane," in *Cold War Crossings: International Travel and Exchange across the Soviet Bloc*, ed. Patryk Babiracki and Kenyon Zimmer (Arlington: University of Texas, 2014), 14–39; "Loopholes in the Iron Curtain" was the title of a conference organised by Philipp Ther and Wlodzimierz Borodziej at the Institut für Osteuropäische Geschichte der Universität Wien in 2013.

14 See Daniel Thomas, "Human Rights Ideas, the Demise of Communism, and the End of the Cold War," *Journal of Cold War Studies* 7, no. 2 (2005): 110–41; Wenger, Mastny and Nuenlist, *Origins*; Adam Roberts, "An 'Incredibly Swift Transition': Reflections on the End of the Cold War," in *Cambridge History of the Cold War*, ed. Melvyn Leffler and Odd Arne Westad (Cambridge: Cambridge University Press, 2010), iii. 513–34; Rosemary Foot, "The Cold War and Human Rights," in *Cambridge History of the Cold War*, ed. Melvyn Leffler and Odd Arne Westad (Cambridge: Cambridge University Press, 2010), iii. 445–65.

15 Angela Romano, "The Conference on Security and Cooperation in Europe: A Reappraisal," in *The Routledge Handbook of the Cold War*, ed. Artemy M. Kalinovsky and Craig Daigle (London: Routledge, 2014), 223–34.

16 Mary E. Sarotte, *Dealing with the Devil: East Germany, détente, and Ostpolitik, 1969–1973* (Chapel Hill: University of North Carolina Press, 2001); Carla Meneguzzi Rostagni, *The Helsinki Process: A Historical Reappraisal* (Padua: Cedam, 2005); Bange and Niedhart, *Helsinki 75*; Wanda Jarzabek, "Hope and Reality: Poland and the Conference on Security and Cooperation in Europe, 1964–1989," *CWIHP Working Paper Series*, no. 56 (2008); Wenger, Mastny, and Nuenlist, *Origins*; Vladimir Bilandzic, Dittar Dahlmann, and Milan Kosanovic, eds., *From Helsinki to Belgrade: The First CSCE Follow-up Meeting and the Crisis of Détente* (Göttingen: Vandenhoeck & Ruprecht, 2012).

17 Angela Romano and Federico Romero, eds., "European Socialist Regimes Facing Globalisation and European Cooperation: Dilemmas and Responses," special issue, *European Review of History* 21, no. 2 (2014).

18 The 5-year-long project *Looking West: The European Socialist regimes facing pan-European cooperation and the European Community* (or *PanEur1970s*) started in 2015 at the European University Institute and received funding from the European Research Council under the European Union's Horizon 2020 research and innovation programme (Grant Agreement n. 669194), https://paneur1970s.eui.eu/.

19 See, for example, Jens Gieseke, *Mielke-Konzern: Die Geschichte der Stasi* (Stuttgart: Deutsche Verlags Anstalt, 2001).

20 Romano, *From Détente*.

21 "CSCE – Economic and Technical Cooperation," NATO Economic Preparation for CSCE: Fiche N.19, Fostering of East–West Travel, 22 December 1972, FCO 28/1705, The National Archives (TNA), London, UK.

22 Luigi Vittorio Ferraris, *Testimonianze di un negoziato: Helsinki, Ginevra, Helsinki 1972–75* (Padua: Cedam, 1977), 438–40. The participating States affirmed their intention to "deal in a positive spirit with questions connected with the allocation of financial means for tourist travel abroad, having regard to their economic possibilities, as well as with those connected with the formalities required for such travel."

23 John Maresca, *To Helsinki: The Conference on Security and Cooperation in Europe 1973–1975* (Durham: Duke University Press, 1985), 178.

24 Victor-Yves Ghebali, *La Diplomatie de la Détente: La CSCE, d'Helsinki à Vienne (1973–1989)* (Brussels: Bruylant, 1989), 246.

25 *Conference on Security and Co-operation in Europe Final Act*, Helsinki 1975, www.osce.org/helsinki-final-act.

26 Mr Bullard (Helsinki) to Mr Wiggin, CSCE: Soviet Attitude, Helsinki, 5 July 1973, Doc. No. 40, in G. Bennett and Keith Hamilton, eds., *Documents on British Policy Overseas: The Conference of Security and Cooperation in Europe 1972–1975* (London: Whitehall History, 1997) (hereafter *DBPO*), 3:ii.

27 Ghebali, *Diplomatie de la Détente*, 274.

28 Ibid., 288–89.

29 Mr Elliott (Geneva) to Sir A. Douglas-Home, Geneva, 15 December 1973, Doc. No. 57, *DBPO*, 218.

30 Minute from Mr Tickell on CSCE: Basket III, FCO, 15 March 1974, Doc. No. 68, *DBPO*.

31 Rapport du Sous-Comité CSCE et du Groupe Ad Hoc au Comité Politique, 17 Janvier 1975, FD 136, Historical Archives of the European Union (hereafter HAEU), Florence, Italy; Sir D. Hildyard (UKMis Geneva) to Mr Callaghan, Geneva, 28 February 1975, Doc. No. 114, *DBPO*.

32 Minute from Br Burns to Mr Tickell, FCO, 29 April 1975, Doc. No. 120, *DBPO*.

33 Minute from Mr Tickell on CSCE: Basket III, FCO, 16 May 1975, Doc. No. 122, *DBPO*.

34 Final Act, Cooperation in Humanitarian and Other Fields, Human Contacts, Preamble, par. 5.

35 Mr Callaghan to HM Representatives Overseas, FCO, 28 July 1975, "Conference on Security and Co-operation (CSCE)," Doc. No. 137, *DBPO*, p. 458.

36 At the 1994 meeting in Budapest the participating countries agreed to turn the CSCE into the Organization for Security and Cooperation in Europe (OSCE).

37 Ghebali, *Diplomatie de la Détente*, 304.

38 Dariusz Stola, *Kraj bez wyjścia: Migracje z Polski 1949–1989* (Warsaw: Instytut Studiów Politycznych PAN, 2010). I would like to thank Aleksandra Kormonicka for bringing the Polish passport policy and book to my attention.

39 György Péteri, "The Occident Within – or the Drive for Exceptionalism and Modernity," *Kritika: Explorations in Russian and Eurasian History* 9, no. 4 (2008): 929–37.

40 Ibid., 936.

41 Paulina Bren, "Mirror, Mirror, on the Wall . . . Is the West the Fairest of Them All? Czechoslovak Normalization and Its (Dis)Contents," *Kritika: Explorations in Russian and Eurasian History* 9, no. 4 (2008): 831–54.

42 Péteri, "Nylon Curtain," 115.

43 Arnd Bauerkämper, "Europe as Social Practice: Towards an Interactive Approach to Modern European History," *East Central Europe* 36 (2009): 30.

References

Bagnato, Bruna. *Prove di Ostpolitik: Politica ed economia nella strategia italiana verso l'Unione Sovietica: 1958–1963*. Florence: L. S. Olschki, 2003.

Bange, Oliver, and Gottfried Niedhart, eds. *Helsinki 75 and the Transformation of Europe*. New York: Berghahn, 2008.

———, and Poul Villaume, eds. *The Long Détente: Changing Concepts of Security and Cooperation in Europe, 1950s–1980s*. Budapest: Central University Press, 2017.

Bauerkämper, Arnd. "Europe as Social Practice: Towards an Interactive Approach to Modern European History." *East Central Europe* 36 (2009): 20–36.

Bennett, G., and K. Hamilton, eds. *Documents on British Policy Overseas: The Conference of Security and Cooperation in Europe 1972–1975*, 3:ii. London: Whitehall History, 1997.

Berend, Ivan T. "What Is Central and Eastern Europe?" *European Journal of Social Theory* 8, no. 4 (2005): 401–16.

Bilandzic, Vladimir, Dittar Dahlmann, and Milan Kosanovic, eds. *From Helsinki to Belgrade: The First CSCE Follow-up Meeting and the Crisis of Détente*. Göttingen: Vandenhoeck & Ruprecht, 2012.

Bozo, Frédéric, Marie-Pierre Rey, N. Piers Ludlow, and Leopoldo Nuti, eds. *Europe and the End of the Cold War: A Reappraisal*. London: Routledge, 2008.

———, and Bernd Rother, eds. *Overcoming the Iron Curtain: Visions of the End of the Cold War in Europe, 1945–1990*. New York: Berghahn, 2012.

Bren, Paulina. "Mirror, Mirror, on the Wall . . . Is the West the Fairest of Them All? Czechoslovak Normalization and Its (Dis)Contents." *Kritika: Explorations in Russian and Eurasian History* 9, no. 4 (2008): 831–54.

David-Fox, Michael. "The Iron Curtain as a Semipermeable Membrane." In *Cold War Crossings: International Travel and Exchange across the Soviet Bloc*, edited by Patryk Babiracki and Kenyon Zimmer, 14–39. Arlington: University of Texas, 2014.

Enderle-Burcel, Gertrude, ed. *Gaps in the Iron Curtain: Economic Relations between Neutral and Socialist Countries in Cold War Europe*. Cracow: Jagiellonian University Press, 2009.

English Robert D. *Russia and the Idea of the West: Gorbachev, Intellectuals, and the End of the Cold War*. New York: Columbia University Press, 2000.

Ferraris, Luigi Vittorio. *Testimonianze di un negoziato: Helsinki, Ginevra, Helsinki 1972–75*. Padua: Cedam, 1977.

Foot, Rosemary. "The Cold War and Human Rights." In *Cambridge History of the Cold War*, edited by Melvyn Leffler and Odd Arne Westad, iii. 445–65. Cambridge: Cambridge University Press, 2010.

Ghebali, Victor-Yves. *La Diplomatie de la Détente: La CSCE, d'Helsinki à Vienne (1973–1989)*. Brussels: Bruylant, 1989.

Gieseke, Jens. *Mielke-Konzern: Die Geschichte der Stasi*. Stuttgart: Deutsche Verlags Anstalt, 2001.

Hanhimäki, Jussi M. "Détente in Europe, 1969–1975." In *Cambridge History of the Cold War*, edited by Melvyn Leffler and Odd Arne Westad, ii. 198–218. Cambridge: Cambridge University Press, 2010.

Hofmann, Arne. *The Emergence of Détente in Europe: Brandt, Kennedy and the Formation of Ostpolitik*. London: Routledge, 2007.

Jarzabek, Wanda. "Hope and Reality: Poland and the Conference on Security and Cooperation in Europe, 1964–1989." *CWIHP Working Paper Series*, no. 56 (2008).

Knipping, Franz, and Matthias Schönwald, eds. *Aufbruch zum Europa der Zweiten Generation. Die europäische Einigung 1969–1984*. Trier: Wissenschaftlicher Verlag Trier, 2004.

Leffler, Melvyn P. *For the Soul of Mankind: The United States, the Soviet Union, and the Cold War*. New York: Hill and Wang, 2007.

Loth, Wilfried. *Overcoming the Cold War: A History of Détente, 1950–1991*. London: Palgrave, 2002.

———, and George-Henri Soutou, eds. *The Making of Détente: Eastern and Western Europe in the Cold War, 1965–75*. London: Routledge, 2008.

Ludlow, N. Piers, ed. *European Integration and the Cold War: Ostpolitik – Westpolitik, 1965–1973*. London: Routledge, 2007.

———. "European Integration and the Cold War." In *Cambridge History of the Cold War*, edited by Melvyn Leffler and Odd Arne Westad, ii. 179–97. Cambridge: Cambridge University Press, 2010.

Maresca, John. *To Helsinki: The Conference on Security and Cooperation in Europe 1973–1975*. Durham: Duke University Press, 1985.

Meneguzzi Rostagni, Carla, ed. *The Helsinki Process: A Historical Reappraisal*. Padua: Cedam, 2005.

Péteri, Gyorgy. "Nylon Curtain: Transnational and Transsystemic Tendencies in the Cultural Life of State-Socialist Russia and East-Central Europe." *Slavonica* 10, no. 2 (2004): 113–23.

———. "The Occident Within – or the Drive for Exceptionalism and Modernity." *Kritika: Explorations in Russian and Eurasian History* 9, no. 4 (2008): 929–37.

Pons, Silvio, and Federico Romero, eds. *Reinterpreting the end of the Cold War: Issues, Interpretations, Periodizations*. London: Frank Cass, 2011.

Rey, Marie-Pierre. *La tentation du rapprochement: France et URSS à l'heure de la détente (1964–1974)*. Paris: Publications de la Sorbonne, 1991.

Richmond, Yale. *Cultural Exchanges and the Cold War: Raising the Iron Curtain*. University Park: Penn State University Press, 2003.

Roberts, Adam. "An 'Incredibly Swift Transition': Reflections on the end of the Cold War." In *Cambridge History of the Cold War*, edited by Melvyn Leffler and Odd Arne Westad, iii. 513–34. Cambridge: Cambridge University Press, 2010.

Romano, Angela. *From Détente in Europe to European Détente: How the West Shaped the Helsinki CSCE*. Brussels: Peter Lang, 2009.

———. "More Cohesive, Still Divergent: Western Europe, the US and the Madrid CSCE Follow-Up Meeting." In *European Integration and the Atlantic Community in the 1980s*, edited by Kiran K. Patel and Ken Weisbrode, 39–58. Cambridge: Cambridge University Press, 2013.

———, and Federico Romero, eds. "The Conference on Security and Cooperation in Europe: A Reappraisal." In *The Routledge Handbook of the Cold War*, edited by Artemy M. Kalinovsky and Craig Daigle, 223–34. London: Routledge, 2014.

———. "European Socialist Regimes Facing Globalisation and European Cooperation: Dilemmas and Responses." Special issue, *European Review of History* 21, no. 2 (2014).

———. "G-7s, European Councils and East–West Economic Relations, 1975–1982." In *International Summitry and Global Governance: The Rise of the G-7 and the European Council, 1974–1991*, edited by Emmanuel Mourlon-Druol and Federico Romero, 198–222. London: Routledge, 2014.

———. "Untying Cold War Knots: The EEC and Eastern Europe in the Long 1970s." *Cold War History* 14, no. 2 (2014): 153–73.

Romero, Federico. *Storia della guerra fredda: L'ultimo conflitto per l'Europa*. Turin: Einaudi, 2009.

———. "Cold War Historiography at the Crossroads." *Cold War History* 14, no. 4 (2014): 685–703.

Sarotte, Mary E. *Dealing with the Devil: East Germany, détente, and Ostpolitik, 1969–1973*. Chapell Hill: University of North Carolina Press, 2001.

Stola, Dariusz. *Kraj bez wyjścia: Migracje z Polski 1949–1989*. Warsaw: Instytut Studiów Politycznych PAN, 2010.

Thomas, Daniel. "Human Rights Ideas, the Demise of Communism, and the End of the Cold War." *Journal of Cold War Studies* 7, no. 2 (2005): 110–41.

Vaïsse, Maurice. *La Grandeur: Politique étrangère du Général de Gaulle, 1958–1969*. Paris: Fayard, 1998.

Varsori, Antonio. *L'Italia nelle relazioni internazionali dal 1943 al 1992*. Rome: Laterza, 1998.

———, ed. *Alle origini del presente: L'Europa occidentale nella crisi degli anni Settanta*. Milan: Franco Angeli, 2007.

Villaume, Poul, and Odd Arne Westad, eds. *Perforating the Iron Curtain: European Détente, Transatlantic Relations, and the Cold War, 1965–1985*. Copenhagen: Museum Tusculanum, 2010.

Wallace, William V., and Roger A. Clarke. *Comecon, Trade, and the West*. London: Frances Pinter, 1986.

Wenger, Andreas, Vojtech Mastny, and Christian Nuenlist, eds. *Origins of the European Security System: The Helsinki Process Revisited, 1965–75*. London: Routledge, 2009.

Westad, Odd Arne, ed. *Reviewing the Cold War: Approaches, Interpretations, and Theories*. London: Frank Cass, 2000.

———. *The Cold War: A World History*. New York: Basic, 2017.

Young, John W. "Western Europe and the End of the Cold War, 1979–1989." In *Cambridge History of the Cold War*, edited by Melvyn Leffler and Odd Arne Westad, iii. 288–310. Cambridge: Cambridge University Press, 2010.

Index